The Gauge Block Handbook

by

Ted Doiron and John Beers

Dimensional Metrology Group
Precision Engineering Division
National Institute of Standards and Technology

The Gauge Block Handbook

by

Ted Doiron and John Beers

Dimensional Metrology Group
Precision Engineering Division
National Institute of Standards and Technology

Preface

The Dimensional Metrology Group, and its predecessors at the National Institute of Standards and Technology (formerly the National Bureau of Standards) have been involved in documenting the science of gauge block calibration almost continuously since the seminal work of Peters and Boyd in 1926 [1]. Unfortunately, most of this documentation has been in the form of reports and other internal documents that are difficult for the interested metrologist outside the Institute to obtain.

On the occasion of the latest major revision of our calibration procedures we decided to assemble and extend the existing documentation of the NIST gauge block calibration program into one document. We use the word assemble rather than write because most of the techniques described have been documented by various members of the Dimensional Metrology Group over the last 20 years. Unfortunately, much of the work is spread over multiple documents, many of the details of the measurement process have changed since the publications were written, and many large gaps in coverage exist. It is our hope that this handbook has assembled the best of the previous documentation and extended the coverage to completely describe the current gauge block calibration process.

Many of the sections are based on previous documents since very little could be added in coverage. In particular, the entire discussion of single wavelength interferometry is due to John Beers [2]; the section on preparation of gauge blocks is due to Clyde Tucker [3]; the section on the mechanical comparator techniques is predominantly from Beers and Tucker [4]; and the appendix on drift eliminating designs is an adaptation for dimensional calibrations of the work of Joseph Cameron [5] on weighing designs. They have, however, been rewritten to make the handbook consistent in style and coverage. The measurement assurance program has been extensively modified over the last 10 years by one of the authors (TD), and chapter 4 reflects these changes.

We would like to thank Mr. Ralph Veale, Mr. John Stoup, Mrs. Trish Snoots, Mr. Eric Stanfield, Mr. Dennis Everett, Mr. Jay Zimmerman, Ms. Kelly Warfield and Dr. Jack Stone, the members of the Dimensional Metrology Group who have assisted in both the development and testing of the current gauge block calibration system and the production of this document.

TD and JSB

CONTENTS

	Page
Preface	1
Introduction	6

1. Length

    1.1   The meter 7

    1.2   The inch 8

2. Gauge blocks

    2.1   A short history of gauge blocks 8

    2.2   Gauge block standards (U.S.) 9

        2.2.1   Scope 9

        2.2.2   Nomenclature and definitions 10

        2.2.3   Tolerance grades 12

        2.2.4   Recalibration requirements 14

    2.3   International standards 15

3. Physical and thermal properties of gauge blocks

    3.1   Materials 17

    3.2   Flatness and parallelism

        3.2.1   Flatness measurement 18

        3.2.2   Parallelism measurement 19

    3.3   Thermal expansion 23

        3.3.1   Thermal expansion of gauge block materials 23

    3.3.2   Thermal expansion uncertainty 27

    3.3   Elastic properties 29

        3.4.1   Contact deformation in mechanical comparisons 30

        3.4.2   Measurement of probe force and tip radius 32

    3.5   Stability 34

4. Measurement assurance programs

- 4.1 Introduction 36
- 4.2 A comparison: traditional metrology vs measurement assurance programs
  - 4.2.1 Tradition 36
  - 4.2.2 Process control: a paradigm shift 37
  - 4.2.3 Measurement assurance: building a measurement process model 39
- 4.3 Determining Uncertainty
  - 4.3.1 Stability 40
  - 4.3.2 Uncertainty 40
  - 4.3.3 Random error 43
  - 4.3.4 Systematic error and type B uncertainty 43
  - 4.3.5 Error budgets 45
  - 4.3.6 Combining type A and type B uncertainties 47
  - 4.3.7 Combining random and systematic errors 48
- 4.4 The NIST gauge block measurement assurance program
  - 4.4.1 Establishing interferometric master values 50
  - 4.4.2 The comparison process 53
    - 4.4.2.1 Measurement schemes - drift eliminating designs ... 54
    - 4.4.2.2 Control parameter for repeatability 57
    - 4.4.2.3 Control test for variance 59
    - 4.4.2.4 Control parameter (S-C) 60
    - 4.4.2.5 Control test for (S-C), the check standard 63
    - 4.4.2.6 Control test for drift 64
  - 4.4.3 Calculating total uncertainty 64
- 4.5 Summary of the NIST measurement assurance program 66

5. The NIST mechanical comparison procedure
- 5.1 Introduction 68
- 5.2 Preparation and inspection 68
  - 5.2.1 Cleaning procedures 68
  - 5.2.2 Cleaning interval 69
  - 5.2.3 Storage 69
  - 5.2.4 Deburring gauge blocks 69
- 5.3 The comparative principle 70
  - 5.3.1 Examples 71
- 5.4 Gauge block comparators 73
  - 5.4.1 Scale and contact force control 75
  - 5.4.2 Stylus force and penetration corrections 75
  - 5.4.3 Environmental factors 77
    - 5.4.3.1 Temperature effects 77
    - 5.4.3.2 Control of temperature effects 79
- 5.5 Intercomparison procedures 80
  - 5.5.1 Handling techniques 81
- 5.6 Comparison designs 82
  - 5.6.1 Drift eliminating designs 82
    - 5.6.1.1 The 12/4 design 82
    - 5.6.1.2 The 6/3 design 83
    - 5.6.1.3 The 8/4 design 84
    - 5.6.1.4 The ABBA design 84
  - 5.6.2 Example of calibration output using the 12/4 design 85
- 5.7 Current NIST system performance 87
  - 5.7.1 Summary 89

6. Gauge block interferometry
- 6.1 Introduction 90
- 6.2 Interferometers 90
  - 6.2.1 The Kosters type interferometer 91

		6.2.2	The NPL interferometer	93
		6.2.3	Testing optical quality of interferometers	95
		6.2.4	Interferometer corrections	96
		6.2.5	Laser light sources	99
	6.3		Environmental conditions and their measurement	99
		6.3.1	Temperature	100
		6.3.2	Atmospheric pressure	101
		6.3.3	Water vapor	101
	6.4		Gauge block measurement procedure	102
	6.5		Computation of gauge block length	104
		6.5.1	Calculation of the wavelength	104
		6.5.2	Calculation of the whole number of fringes	105
		6.5.3	Calculation of the block length from observed data	106
	6.6		Type A and B errors	107
	6.7		Process evaluation	109
	6.8		Multiple wavelength interferometry	112
	6.9		Use of the line scale interferometer for end standard calibration	114
7. References				117

Appendix A.	Drift eliminating designs for non-simultaneous comparison calibrations	121
Appendix B.	Wringing films	134
Appendix C.	Phase shifts in gauge block interferometry	137
Appendix D.	Deformation corrections	141

Gauge Block Handbook

Introduction

Gauge block calibration is one of the oldest high precision calibrations made in dimensional metrology. Since their invention at the turn of the century gauge blocks have been the major source of length standardization for industry. In most measurements of such enduring importance it is to be expected that the measurement would become much more accurate and sophisticated over 80 years of development. Because of the extreme simplicity of gauge blocks this has only been partly true. The most accurate measurements of gauge blocks have not changed appreciably in accuracy in the last 70 years. What has changed is the much more widespread necessity of such accuracy. Measurements, which previously could only be made with the equipment and expertise of a national metrology laboratory, are routinely expected in private industrial laboratories.

To meet this widespread need for higher accuracy, the calibration methods used for gauge blocks have been continuously upgraded. This handbook is a both a description of the current practice at the National Institute of Standards and Technology, and a compilation of the theory and lore of gauge block calibration. Most of the chapters are nearly self-contained so that the interested reader can, for example, get information on the cleaning and handling of gauge blocks without having to read the chapters on measurement schemes or process control, etc. This partitioning of the material has led to some unavoidable repetition of material between chapters.

The basic structure of the handbook is from the theoretical to the practical. Chapter 1 concerns the basic concepts and definitions of length and units. Chapter 2 contains a short history of gauge blocks, appropriate definitions and a discussion of pertinent national and international standards. Chapter 3 discusses the physical characteristics of gauge blocks, including thermal, mechanical and optical properties. Chapter 4 is a description of statistical process control (SPC) and measurement assurance (MA) concepts. The general concepts are followed by details of the SPC and MA used at NIST on gauge blocks.

Chapters 5 and 6 cover the details of the mechanical comparisons and interferometric techniques used for gauge block calibrations. Full discussions of the related uncertainties and corrections are included. Finally, the appendices cover in more detail some important topics in metrology and gauge block calibration.

1.0 Length

1.1 The Meter

At the turn of 19th century there were two distinct major length systems. The metric length unit was the meter that was originally defined as 1/10,000,000 of the great arc from the pole to the equator, through Paris. Data from a very precise measurement of part of that great arc was used to define an artifact meter bar, which became the practical and later legal definition of the meter. The English system of units was based on a yard bar, another artifact standard [6].

These artifact standards were used for over 150 years. The problem with an artifact standard for length is that nearly all materials are slightly unstable and change length with time. For example, by repeated measurements it was found that the British yard standard was slightly unstable. The consequence of this instability was that the British inch (1/36 yard) shrank [7], as shown in table **1.1**.

Table 1.1

Year	Value
1895 -	25.399978 mm
1922 -	25.399956 mm
1932 -	25.399950 mm
1947 -	25.399931 mm

The first step toward replacing the artifact meter was taken by Albert Michelson, at the request of the International Committee of Weights and Measures (CIPM). In 1892 Michelson measured the meter in terms of the wavelength of red light emitted by cadmium. This wavelength was chosen because it has high coherence, that is, it will form fringes over a reasonable distance. Despite the work of Michelson, the artifact standard was kept until 1960 when the meter was finally redefined in terms of the wavelength of light, specifically the red-orange light emitted by excited krypton-86 gas.

Even as this definition was accepted, the newly invented helium-neon laser was beginning to be used for interferometry. By the 1970's a number of wavelengths of stabilized lasers were considered much better sources of light than krypton red-orange for the definition of the meter. Since there were a number of equally qualified candidates the International Committee on Weights and Measures (CIPM) decided not to use any particular wavelength, but to make a change in the measurement hierarchy. The solution was to define the speed of light in vacuum as exactly 299,792,458 m/s, and make length a derived unit. In theory, a meter can be produced by anyone with an accurate clock [8].

In practice, the time-of-flight method is impractical for most measurements, and the meter is measured using known wavelengths of light. The CIPM lists a number of laser and atomic sources and recommended frequencies for the light. Given the defined speed of light, the wavelength of the light can be calculated, and a meter can be generated by counting wavelengths of the light. Methods for this measurement are discussed in the chapter on interferometry.

1.2 The Inch

In 1866, the United Stated Surveyor General decided to base all geodetic measurements on an inch defined from the international meter. This inch was defined such that there were exactly 39.37 inches in the meter. England continued to use the yard bar to define the inch. These different inches continued to coexist for nearly 100 years until quality control problems during World War II showed that the various inches in use were too different for completely interchangeable parts from the English speaking nations. Meetings were held in the 1950's and in 1959 the directors of the national metrology laboratories of the United States, Canada, England, Australia and South Africa agreed to define the inch as 25.4 millimeters, exactly [9]. This definition was a compromise; the English inch being somewhat longer, and the U.S. inch smaller. The old U.S. inch is still in use for commercial surveying of land in the form of the "surveyor's foot," which is 12 old U.S. inches.

2.0 Gauge Blocks

2.1 A Short History of Gauge Blocks

By the end of the nineteenth century the idea of interchangeable parts begun by Eli Whitney had been accepted by industrial nations as the model for industrial manufacturing. One of the drawbacks to this new system was that in order to control the size of parts numerous gauges were needed to check the parts and set the calibrations of measuring instruments. The number of gauges needed for complex products, and the effort needed to make and maintain the gauges was a significant expense. The major step toward simplifying this situation was made by C.E. Johannson, a Swedish machinist.

Johannson's idea, first formulated in 1896 [10], was that a small set of gauges that could be combined to form composite gauges could reduce the number of gauges needed in the shop. For example, if four gauges of sizes 1 mm, 2 mm, 4 mm, and 8 mm could be combined in any combination, all of the millimeter sizes from 1 mm to 15 mm could be made from only these four gauges. Johannson found that if two opposite faces of a piece of steel were lapped very flat and parallel, two blocks would stick together when they were slid together with a very small amount of grease between them. The width of this "wringing" layer is about 25 nm, and was so small for the tolerances needed at the time, that the block lengths could be added together with no correction for interface thickness. Eventually the wringing layer was defined as part of the length of the block, allowing the use of an unlimited number of wrings without correction for the size of the wringing layer.

In the United States, the idea was enthusiastically adopted by Henry Ford, and from his example the use of gauge blocks was eventually adopted as the primary transfer standard for length in industry. By the beginning of World War I, the gauge block was already so important to industry that the Federal Government had to take steps to insure the availability of blocks. At the outbreak of the war, the only supply of gauge blocks was from Europe, and this supply was interrupted.

In 1917 inventor William Hoke came to NBS proposing a method to manufacture gauge blocks equivalent to those of Johannson [11]. Funds were obtained from the Ordnance Department for the project and 50 sets of 81 blocks each were made at NBS. These blocks were cylindrical and had a hole in the center, the hole being the most prominent feature of the design. The current generation of square cross-section blocks have this hole and are referred to as "Hoke blocks."

2.2 Gauge Block Standards (U.S.)

There are two main American standards for gauge blocks, the Federal Specification GGG-G-15C [12] and the American National Standard ANSI/ASME B89.1.9M [13]. There are very few differences between these standards, the major ones being the organization of the material and the listing of standard sets of blocks given in the GGG-G-15C specification. The material in the ASME specification that is pertinent to a discussion of calibration is summarized below.

2.2.1 Scope

The ASME standard defines all of the relevant physical properties of gauge blocks up to 20 inches and 500 mm long. The properties include the block geometry (length, parallelism, flatness and surface finish), standard nominal lengths, and a tolerance grade system for classifying the accuracy level of blocks and sets of blocks.

The tolerancing system was invented as a way to simplify the use of blocks. For example, suppose gauge blocks are used to calibrate a certain size fixed gauge, and the required accuracy of the gauge is 0.5 μm. If the size of the gauge requires a stack of five blocks to make up the nominal size of the gauge the accuracy of each block must be known to 0.5/5 or 0.1 μm. This is near the average accuracy of an industrial gauge block calibration, and the tolerance could be made with any length gauge blocks if the calibrated lengths were used to calculate the length of the stack. But having the calibration report for the gauge blocks on hand and calculating the length of the block stack are a nuisance. Suppose we have a set of blocks which are guaranteed to have the property that each block is within 0.05 μm of its nominal length. With this knowledge we can use the blocks, assume the nominal lengths and still be accurate enough for the measurement.

The tolerance grades are defined in detail in section 2.2.3, but it is important to recognize the difference between gauge block calibration and certification. At NIST, gauge blocks are calibrated, that is, the measured length of each block is reported in the calibration report. The report does not state which tolerance grade the blocks satisfy. In many industrial calibrations only the certified tolerance grade is reported since the corrections will not be used.

2.2.2 Nomenclature and Definitions

A gauge block is a length standard having flat and parallel opposing surfaces. The cross-sectional shape is not very important, although the standard does give suggested dimensions for rectangular, square and circular cross-sections. Gauge blocks have nominal lengths defined in either the metric system (millimeters) or in the English system (1 inch = 25.4 mm).

The length of the gauge block is defined at standard reference conditions:

>temperature = 20 °C (68 °F)
>barometric pressure = 101,325 Pa (1 atmosphere)
>water vapor pressure = 1,333 Pa (10 mm of mercury)
>CO_2 content of air = 0.03%.

Of these conditions only the temperature has a measurable effect on the physical length of the block. The other conditions are needed because the primary measurement of gauge block length is a comparison with the wavelength of light. For standard light sources the frequency of the light is constant, but the wavelength is dependent on the temperature, pressure, humidity, and CO_2 content of the air. These effects are described in detail later.

The length of a gauge block is defined as the perpendicular distance from a gauging point on one end of the block to an auxiliary true plane wrung to the other end of the block, as shown in figure 2.1 (from B89.1.9).

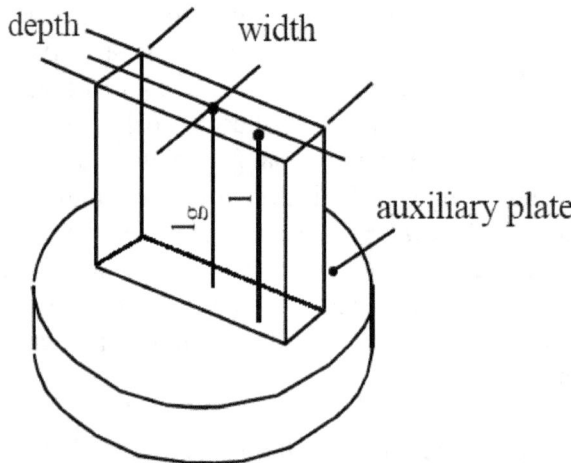

Figure 2.1. The length of a gauge block is the distance from the gauging point on the top surface to the plane of the platen adjacent to the wrung gauge block.

This length is measured interferometrically, as described later, and corrected to standard conditions. It is worth noting that gauge blocks are NEVER measured at standard conditions because the standard vapor pressure of water of 10 mm of mercury is nearly 60% relative humidity that would allow steel to rust. The standard conditions are actually spectroscopic standard conditions, i.e., the conditions at which spectroscopists define the wavelengths of light.

This definition of gauge block length that uses a wringing plane seems odd at first, but is very important for two reasons. First, light appears to penetrate slightly into the gauge block surface, a result of the surface finish of the block and the electromagnetic properties of metals. If the wringing plane and the gauge block are made of the same material and have the same surface finish, then the light will penetrate equally into the block top surface and the reference plane, and the errors cancel. If the block length was defined as the distance between the gauge block surfaces the penetration errors would add, not cancel, and the penetration would have to be measured so a correction could be made. These extra measurements would, of course, reduce the accuracy of the calibration.

The second reason is that in actual use gauge blocks are wrung together. Suppose the length of gauge blocks was defined as the actual distance between the two ends of the gauge block, not wrung to a plane. For example, if a length of 6.523 mm is needed gauge blocks of length 2.003 mm, 2.4 mm, and 2.12 mm are wrung together. The length of this stack is 6.523 plus the length of two wringing layers. It could also be made using the set (1 mm, 1 mm, 1 mm, 1.003 mm, 1.4 mm, and 1.12 mm) which would have the length of 6.523 mm plus the length of 5 wringing layers. In order to use the blocks these wringing layer lengths must be known. If, however, the length of each block contains one wringing layer length then both stacks will be of the same defined length.

NIST master gauge blocks are calibrated by interferometry in accordance with the definition of gauge block length. Each master block carries a wringing layer with it, and this wringing layer is transferred to every block calibrated at NIST by mechanical comparison techniques.

The mechanical length of a gauge block is the length determined by mechanical comparison of a block to another block of known interferometrically determined length. The mechanical comparison must be a measurement using two designated points, one on each end of the block. Since most gauge block comparators use mechanical contact for the comparison, if the blocks are not of the same material corrections must be made for the deformation of the blocks due to the force of the comparator contact.

The reference points for rectangular blocks are the center points of each gauging face. For square gauge block mechanical comparison are shown in figure 2.2.

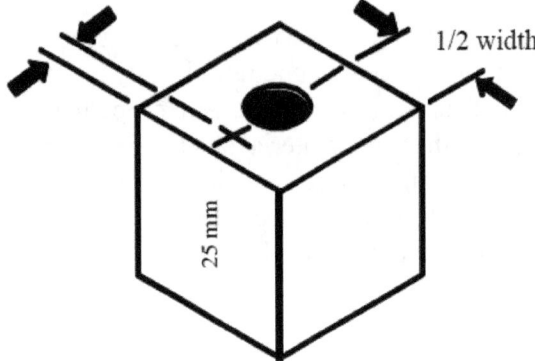

Figure 2.2 Definition of the gauging point on square gauge blocks.

For rectangular and round blocks the reference point is the center of gauging face. For round or square blocks that have a center hole, the point is midway between the hole edge and the edge of the block nearest to the size marking.

2.2.3 Tolerance Grades

There are 4 tolerance grades; 0.5, 1, 2, and 3. Grades 0.5 and 1 gauge blocks have lengths very close to their nominal values. These blocks are generally used as calibration masters. Grades 2 and 3 are of lower quality and are used for measurement and gauging purposes. Table 2.1 shows the length, flatness and parallelism requirements for each grade. The table shows that grade 0.5 blocks are within 1 millionth of an inch (1 μin) of their nominal length, with grades 1, 2, and 3 each roughly doubling the size of the maximum allowed deviation.

Table 2.1a Tolerance Grades for Inch Blocks (in µin)

Nominal	Grade .5	Grade 1	Grade 2	Grade 3
<1 inch	1	2	+4, -2	+8, -4
2	2	4	+8, -4	+16, -8
3	3	5	+10, -5	+20, -10
4	4	6	+12, -6	+24, -12
5		7	+14, -7	+28, -14
6		8	+16, -8	+32, -16
7		9	+18, -9	+36, -18
8		10	+20, -10	+40, -20
10		12	+24, -12	+48, -24
12		14	+28, -14	+56, -28
16		18	+36, -18	+72, -36
20		20	+40, -20	+80, -40

Table 2.1b Tolerance Grades for Metric Blocks (µm)

Nominal	Grade .5	Grade 1	Grade 2	Grade 3
< 10 mm	0.03	0.05	+0.10, -0.05	+0.20, -0.10
< 25 mm	0.03	0.05	+0.10, -0.05	+0.30, -0.15
< 50 mm	0.05	0.10	+0.20, -0.10	+0.40, -0.20
< 75 mm	0.08	0.13	+0.25, -0.13	+0.45, -0.23
< 100 mm	0.10	0.15	+0.30, -0.15	+0.60, -0.30
125 mm		0.18	+0.36, -0.18	+0.70, -0.35
150 mm		0.20	+0.41, -0.20	+0.80, -0.40
175 mm		0.23	+0.46, -0.23	+0.90, -0.45
200 mm		0.25	+0.51, -0.25	+1.00, -0.50
250 mm		0.30	+0.60, -0.30	+1.20, -0.60
300 mm		0.35	+0.70, -0.35	+1.40, -0.70
400 mm		0.45	+0.90, -0.45	+1.80, -0.90
500 mm		0.50	+1.00, -0.50	+2.00, -1.90

Since there is uncertainty in any measurement, the standard allows for an additional tolerance for length, flatness, and parallelism. These additional tolerances are given in table **2.2**.

Table 2.2 Additional Deviations for Measurement Uncertainty

Nominal in (mm)	Grade .5 µin (µm)	Grade 1 µin (µm)	Grade 2 µin (µm)	Grade 3 µin (µm)
< 4 (100)	1 (0.03)	2 (0.05)	3 (0.08)	4 (0.10)
< 8 (200)		3 (0.08)	6 (0.15)	8 (0.20)
< 12 (300)		4 (0.10)	8 (0.20)	10 (0.25)
< 20 (500)		5 (0.13)	10 (0.25)	12 (0.30)

For example, for a grade 1 gauge block of nominally 1 inch length the length tolerance is 2 µin. With the additional tolerance for measurement uncertainty from table 2 of 2 uin, a 1 in grade 1 block must have a measured length within 4 uin of nominal.

2.2.4 Recalibration Requirements

There is no required schedule for recalibration of gauge blocks, but both the ASME and Federal standards recommend recalibration periods for each tolerance grade, as shown below:

Grade	Recalibration Period
0.5	Annually
1	Annually
2	Monthly to semi-annually
3	Monthly to semi-annually

Most NIST customers send master blocks for recalibration every 2 years. Since most master blocks are not used extensively, and are used in a clean, dry environment this schedule is probably adequate. However, despite popular misconceptions, **NIST has no regulatory power in these matters**. The rules for tolerance grades, recalibration, and replacement rest entirely with the appropriate government agency inspectors.

2.3 International Standards

Gauge blocks are defined internationally by ISO Standard 3650 [14], the current edition being the first edition 1978-07-15. This standard is much like the ANSI standard in spirit, but differs in most details, and of course does not define English size blocks.

The length of the gauge block is defined as the distance between a flat surface wrung to one end of the block, and a gauging point on the opposite end. The ISO specification only defines rectangular cross-sectioned blocks and the gauging point is the center of the gauging face. The non-gauging dimensions of the blocks are somewhat smaller than the corresponding ANSI dimensions.

There are four defined tolerance grades in ISO 3650; 00, 0, 1 and 2. The algorithm for the length tolerances are shown in table 2.3, and there are rules for rounding stated to derive the tables included in the standard.

Table 2.3

Grade	Deviation from Nominal Length (μm)
00	(0.05 + 0.0001L)
0	(0.10 + 0.0002L)
1	(0.20 + 0.0004L)
2	(0.40 + 0.0008L)

Where L is the block nominal length in millimeters.

The ISO standard does not have an added tolerance for measurement uncertainty; however, the ISO tolerances are comparable to those of the ANSI specification when the additional ANSI tolerance for measurement uncertainty is added to the tolerances of Table 2.1.

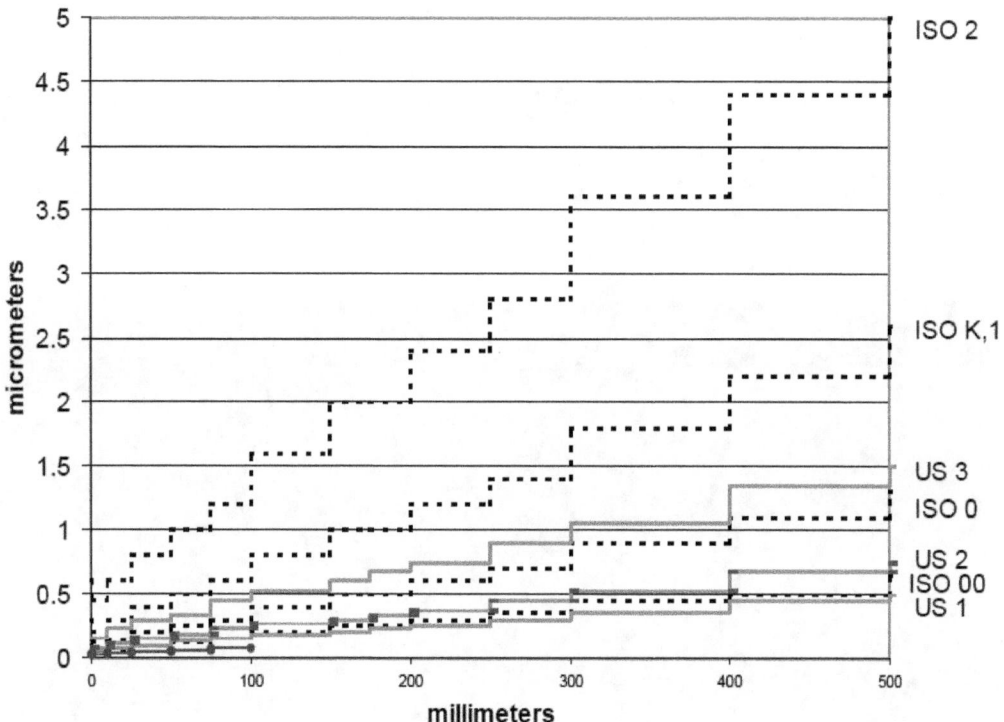

Figure 2.3. Comparison of ISO grade tolerances (black dashed) and ASME grade tolerances (red).

A graph of the length tolerance versus nominal length is shown in figure 2.3. The different class tolerance for ISO and ANSI do not match up directly. The ANSI grade 1 is slightly tighter than ISO class 00, but if the additional ANSI tolerance for measurement uncertainty is used the ISO Grade 00 is slightly tighter. The practical differences between these specifications are negligible.

In many countries the method for testing the variation in length is also standardized. For example, in Germany [15] the test block is measured in 5 places: the center and near each corner (2 mm from each edge). The center gives the length of the block and the four corner measurements are used to calculate the shortest and longest lengths of the block. Some of the newer gauge block comparators have a very small lower contact point to facilitate these measurements very near the edge of the block.

3. Physical and Thermal Properties of Gauge Blocks

3.1 Materials

From the very beginning gauge blocks were made of steel. The lapping process used to finish the ends, and the common uses of blocks demand a hard surface. A second virtue of steel is that most industrial products are made of steel. If the steel gauge block has the same thermal expansion coefficient as the part to be gauged, a thermometer is not needed to obtain accurate measurements. This last point will be discussed in detail later.

The major problem with gauge blocks was always the stability of the material. Because of the hardening process and the crystal structure of the steel used, most blocks changed length in time. For long blocks, over a few inches, the stability was a major limitation. During the 1950s and 1960s a program to study the stability problem was sponsored by the National Bureau of Standards and the ASTM [16,17]. A large number of types of steel and hardening processes were tested to discover manufacturing methods that would produce stable blocks. The current general use of 52100 hardened steel is the product of this research. Length changes of less than 1 part in 10^{-6}/decade are now common.

Over the years, a number of other materials were tried as gauge blocks. Of these, tungsten carbide, chrome carbide, and Cervit are the most interesting cases.

The carbide blocks are very hard and therefore do not scratch easily. The finish of the gauging surfaces is as good as steel, and the lengths appear to be at least as stable as steel, perhaps even more stable. Tungsten carbide has a very low expansion coefficient (1/3 of steel) and because of the high density the blocks are deceptively heavy. Chrome carbide has an intermediate thermal expansion coefficient (2/3 of steel) and is roughly the same density as steel. Carbide blocks have become very popular as master blocks because of their durability and because in a controlled laboratory environment the thermal expansion difference between carbide and steel is easily manageable.

Cervit is a glassy ceramic that was designed to have nearly zero thermal expansion coefficient. This property, plus a zero phase shift on quartz platens (phase shift will be discussed later), made the material attractive for use as master blocks. The drawbacks are that the material is softer than steel, making scratches a danger, and by nature the ceramic is brittle. While a steel block might be damaged by dropping, and may even need stoning or recalibration, Cervit blocks tended to crack or chip. Because the zero coefficient was not always useful and because of the combination of softness and brittleness they never became popular and are no longer manufactured.

A number of companies are experimenting with zirconia based ceramics, and one type is being marketed. These blocks are very hard and have thermal expansion coefficient of approximately 9×10^{-6}/°C, about 20% lower than steel.

3.2 Flatness and Parallelism

We will describe a few methods that are useful to characterize the geometry of gauge blocks. It is important to remember, however, that these methods provide only a limited amount of data about what can be, in some cases, a complex geometric shape. When more precise measurements or a permanent record is needed, the interference fringe patterns can be photographed. The usefulness of each of the methods must be judged in the light of the user's measurement problem.

3.2.1 Flatness Measurements.

Various forms of interferometers are applicable to measuring gauge block flatness. All produce interference fringe patterns formed with monochromatic light by the gauge block face and a reference optical flat of known flatness. Since modest accuracies (25 nm or 1 µin) are generally needed, the demands on the light source are also modest. Generally a fluorescent light with a green filter will suffice as an illumination source. For more demanding accuracies, a laser or atomic spectral lamp must be used.

The reference surface must satisfy two requirements. First, it must be large enough to cover the entire surface of the gauge block. Usually a 70 mm diameter or larger is sufficient. Secondly, the reference surface of the flat should be sufficiently planar that any fringe curvature can be attributed solely to the gauge block. Typical commercially available reference flats, flat to 25 nm over a 70 mm diameter, are usually adequate.

Gauge blocks 2 mm (0.1 in.) and greater can be measured in a free state, that is, not wrung to a platen. Gauge blocks less than 2 mm are generally flexible and have warped surfaces. There is no completely meaningful way to define flatness. One method commonly used to evaluate the "flatness" is by "wringing" the block to another more mechanically stable surface. When the block is wrung to the surface the wrung side will assume the shape of the surface, thus this surface will be as planar as the reference flat.

We wring these thin blocks to a fused silica optical flat so that the wrung surface can be viewed through the back surface of the flat. The interface between the block and flat, if wrung properly, should be a uniform gray color. Any light or colored areas indicate poor wringing contact that will cause erroneous flatness measurements. After satisfactory wringing is achieved the upper (non-wrung) surface is measured for flatness. This process is repeated for the remaining surface of the block.

Figures **3.1a** and **3.1b** illustrate typical fringe patterns. The angle between the reference flat and gauge block is adjusted so that 4 or 5 fringes lie across the width of the face of the block, as in figure **3.1a**, or 2 or 3 fringes lie along the length of the face as in figure **3.1b**. Four fringes in each direction are adequate for square blocks.

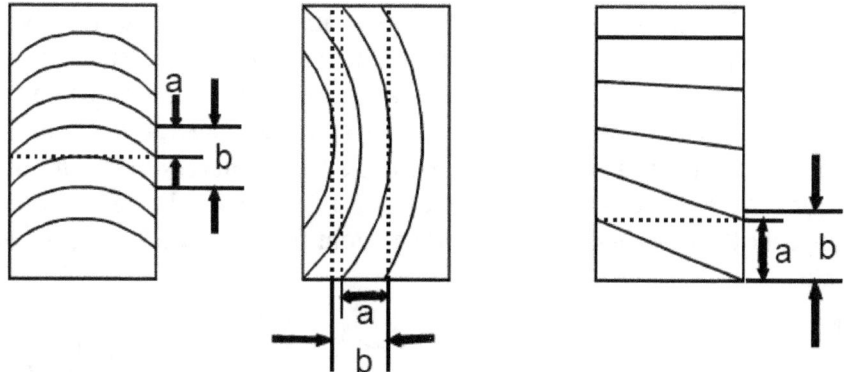

Figure 3.1 a, b, and c. Typical fringe patterns used to measure gauge block flatness. Curvature can be measured as shown in the figures.

The fringe patterns can be interpreted as contour maps. Points along a fringe are points of equal elevation and the amount of fringe curvature is thus a measure of planarity.

$$\text{Curvature} = a/b \quad \text{(in fringes)}.$$

For example, a/b is about 0.2 fringe in figure **3.1a** and 0.6 fringe in figure **3.1b**. Conversion to length units is accomplished using the known wavelength of the light. Each fringe represents a one-half wavelength difference in the distance between the reference flat and the gauge block. Green light is often used for flatness measurements. Light in the green range is approximately 250 nm (10 µin) per fringe, therefore the two illustrations indicate flatness deviations of 50 nm and 150 nm (2 µin and 6 µin) respectively.

Another common fringe configuration is shown in figure **3.1c**. This indicates a twisted gauging face. It can be evaluated by orienting the uppermost fringe parallel to the upper gauge block edge and then measuring "a" and "b" in the two bottom fringes. the magnitude of the twist is a/b which in this case is 75 nm (3 µin) in green.

In manufacturing gauge blocks, the gauging face edges are slightly beveled or rounded to eliminate damaging burrs and sharpness. Allowance should be made for this in flatness measurements by excluding the fringe tips where they drop off at the edge. Allowances vary, but 0.5 mm (0.02 in) is a reasonable bevel width to allow.

3.2.2 Parallelism measurement

Parallelism between the faces of a gauge block can be measured in two ways; with interferometry or with an electro-mechanical gauge block comparator.

Interferometer Technique

The gauge blocks are first wrung to what the standards call an auxiliary surface. We will call these surfaces platens. The platen can be made of any hard material, but are usually steel or glass. An optical flat is positioned above the gauge block, as in the flatness measurement, and the fringe patterns are observed. Figure **3.2** illustrates a typical fringe pattern. The angle between the reference flat and gauge block is adjusted to orient the fringes across the width of the face as in Figure **3.2** or along the length of the face. The reference flat is also adjusted to control the number of fringes, preferably 4 or 5 across, and 2 or 3 along. Four fringes in each direction are satisfactory for square blocks.

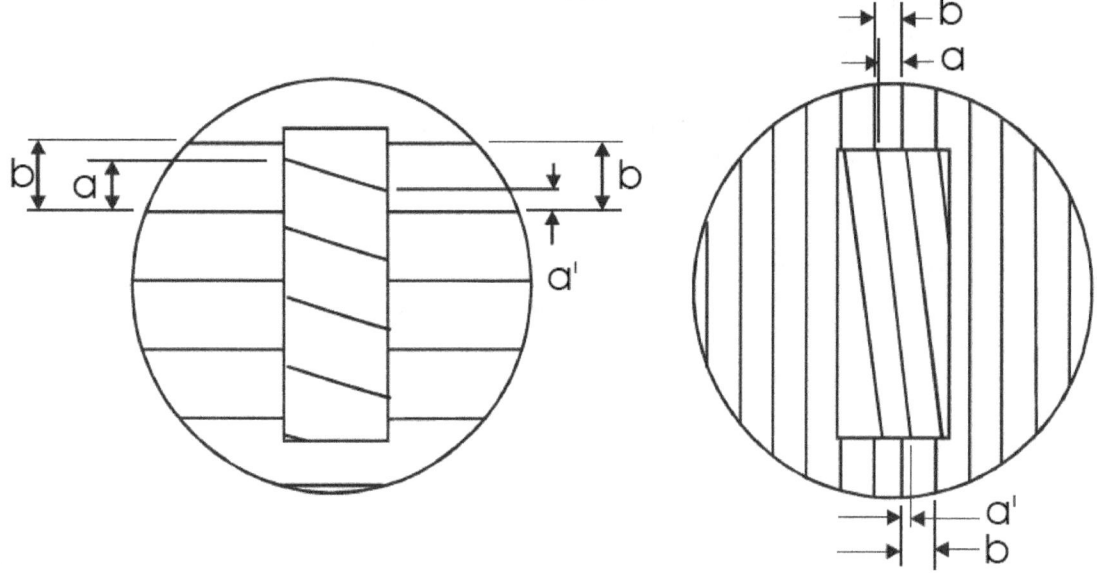

Figure 3.2 Typical fringe patterns for measuring gauge block parallelism using the interferometer method.

A parallelism error between the two faces is indicated by the slope of the gauge block fringes relative to the platen fringes. Parallelism across the block width is illustrated in figure **3.2a** where

$$\text{Slope} = (a/b) - (a'/b) = 0.8 - 0.3 = 0.5 \text{ fringe} \qquad (3.1)$$

Parallelism along the block length in figure **3.2b** is

$$\text{Slope} = (a/b) + (a')/b = 0.8 + 0.3 = 1.1 \text{ fringe} \qquad (3.2)$$

Note that the fringe fractions are subtracted for figure **3.2a** and added for figure **3.2b**. The reason for this is clear from looking at the patterns - the block fringe stays within the same two platen fringes in the first case and it extends into the next pair in the latter case. Conversion to length units is made with the value of $\lambda/2$ appropriate to the illumination.

Since a fringe represents points of equal elevation it is easy to visualize the blocks in figure **3.2** as being slightly wedge shaped.

This method depends on the wringing characteristics of the block. If the wringing is such that the platen represents an extension of the lower surface of the block then the procedure is reliable. There are a number of problems that can cause this method to fail. If there is a burr on the block or platen, if there is particle of dust between the block and platen, or if the block is seriously warped, the entire face of the block may not wring down to the platen properly and a false measurement will result. For this reason usually a fused silica platen is used so that the wring can be examined by looking through the back of the platen, as discussed in the section on flatness measurements. If the wring is good, the block-platen interface will be a fairly homogeneous gray color.

Gauge Block Comparator Technique

Electro-mechanical gauge block comparators with opposing measuring styli can be used to measure parallelism. A gauge block is inserted in the comparator, as shown in figure **3.3**, after sufficient temperature stabilization has occurred to insure that the block is not distorted by internal temperature gradients. Variations in the block thickness from edge to edge in both directions are measured, that is, across the width and along the length of the gauging face through the gauging point. Insulated tongs are recommended for handling the blocks to minimize temperature effects during the measuring procedure.

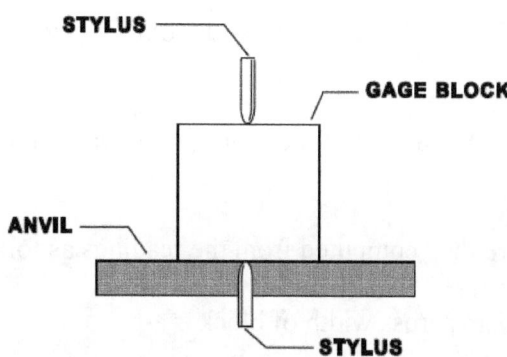

Figure 3.3. Basic geometry of measurements using a mechanical comparator.

Figures **3.4a** and **3.4b** show locations of points to be measured with the comparator on the two principle styles of gauge blocks. The points designated a, b, c, and d are midway along the edges and in from the edge about 0.5 mm (0.02 in) to allow for the normal rounding of the edges.

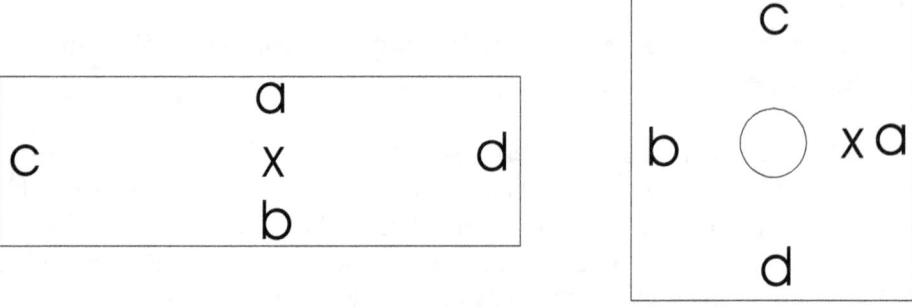

Figure 3.4 a and b. Location of gauging points on gauge blocks for both length (X) and parallelism (a,b,c,d) measurements.

A consistent procedure is recommended for making the measurements:

(1) Face the side of the block associated with point "a" toward the comparator measuring tips, push the block in until the upper tip contacts point "a", record meter reading and withdraw the block.

(2) Rotate the block 180° so the side associated with point "b" faces the measuring tips, push the block in until tip contacts point "b", record meter reading and withdraw block.

(3) Rotate block 90° so side associated with point "c" faces the tips and proceed as in previous steps.

(4) Finally rotate block 180° and follow this procedure to measure at point "d".

The estimates of parallelism are then computed from the readings as follows:

Parallelism across width of block = a-b

Parallelism along length of block = c-d

The parallelism tolerances, as given in the GGG and ANSI standards, are shown in table 3.1.

Table 3.1 ANSI tolerances for parallelism in microinches

Size (in)	Grade .5	Grade 1	Grade 2	Grade 3
<1	1	2	4	5
2	1	2	4	5
3	1	3	4	5
4	1	3	4	5
5-8		3	4	5
10-20		4	5	6

Referring back to the length tolerance table, you will see that the allowed parallelism and flatness errors are very substantial for blocks under 25 mm (or 1 in). For both interferometry and mechanical comparisons, if measurements are made with little attention to the true gauge point significant errors can result when large parallelism errors exist.

3.3 Thermal Expansion

In most materials, a change in temperature causes a change in dimensions. This change depends on both the size of the temperature change and the temperature at which the change occurs. The equation describing this effect is

$$\Delta L/L = \alpha_L \Delta T \tag{3.3}$$

where L is the length, ΔL is the change in length of the object, ΔT is the temperature change and α_L is the coefficient of thermal expansion(CTE).

3.3.1 Thermal Expansion of Gauge Block Materials

In the simplest case, where ΔT is small, α_L can be considered a constant. In truth, α_L depends on the absolute temperature of the material. Figure 3.5 [18] shows the measured expansion coefficient of gauge block steel. This diagram is typical of most metals, the thermal expansion rises with temperature.

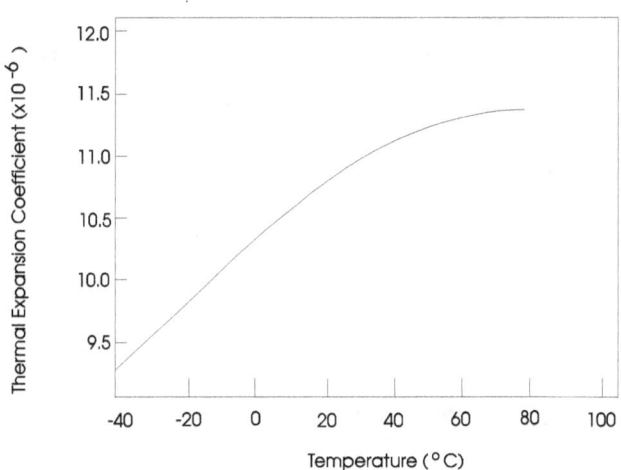

Figure 3.5. Variation of the thermal expansion coefficient of gauge block steel with temperature.

As a numerical example, gauge block steel has an expansion coefficient of $11.5 \times 10^{-6}/°C$. This means that a 100 mm gauge block will grow 11.5×10^{-6} times 100 mm, or 1.15 micrometer, when its temperature is raised 1 °C. This is a significant change in length, since even class 3 blocks are expected to be within 0.2 μm of nominal. For long standards the temperature effects can be dramatic. Working backwards, to produce a 0.25 μm change in a 500 mm gauge block, a temperature change of only 43 millidegrees (0.043 °C) is needed.

Despite the large thermal expansion coefficient, steel has always been the material of choice for gauge blocks. The reason for this is that most measuring and manufacturing machines are made of steel, and the thermal effects tend to cancel.

To see how this is true, suppose we wish to have a cube made in the shop, with a side length of 100 mm. The first question to be answered is at what temperature should the length be 100 mm. As we have seen, the dimension of most objects depends on its temperature, and therefore a dimension without a defined temperature is meaningless. For dimensional measurements the standard temperature is 20 °C (68 °F). If we call for a 100 mm cube, what we want is a cube which at 20 °C will measure 100 mm on a side.

Suppose the shop floor is at 25 °C and we have a perfect gauge block with zero thermal expansion coefficient. If we make the cube so that each side is exactly the same length as the gauge block, what length is it? When the cube is taken into the metrology lab at 20 °C, it will shrink 11.5×10^{-6} /°C, which for our block is 5.75 μm, i.e., it will be 5.75 μm undersized.

Now suppose we had used a steel gauge block. When we brought the gauge block out onto the shop floor it would have grown 5.75 μm. The cube, being made to the dimension of the gauge block would have been oversized by 5.75 μm. And finally, when the block and cube were brought into the gauge lab they would both shrink the same amount, 5.75 μm, and be exactly the length called for in the specification.

What this points out is that the difference in thermal expansion between the workpiece and the gauge is the important parameter. Ideally, when making brass or aluminum parts, brass or aluminum gauges would be used. This is impractical for a number of reasons, not the least of which is that it is nearly impossible to make gauge blocks out of soft materials, and once made the surface would be so easily damaged that its working life would be on the order of days. Another reason is that most machined parts are made from steel. This was particularly true in the first half of the century when gauge blocks were invented because aluminum and plastic were still undeveloped technologies.

Finally, the steel gauge block can be used to gauge any material if corrections are made for the differential thermal expansion of the two materials involved. If a steel gauge block is used to gauge a 100 mm aluminum part at 25 °C, a correction factor must be used. Since the expansion coefficient of aluminum is about twice that of steel, when the part is brought to 20 °C it will shrink twice as much as the steel. Thus the aluminum block must be made oversized by the amount

$$\Delta L = (\alpha_L^{aluminum} - \alpha_L^{steel}) \times L \times \Delta T \tag{3.4}$$

$$= (24 - 11.5) \times 10^{-6} \times 100 \text{ mm} \times 5°C$$

$$= 6.25 \text{ μm}$$

So if we make the cube 6.25 μm larger than the steel gauge block it will be exactly 100 mm when brought to standard conditions (20 °C).

There are a few mixtures of materials, alloys such as invar and crystalline/glass mixtures such as Zerodur, which have small thermal expansion coefficients. These materials are a combination of two components, one which expands with increasing temperature, and one which shrinks with increasing temperature. A mixture is made for which the expansion of one component matches the shrinkage of the other. Since the two materials never have truly opposite temperature dependencies, the matching of expansion and shrinkage can be made at only one temperature. It is important to remember that these materials are designed for one temperature, usually 20 °C, and the thermal expansion coefficient can be much different even a few degrees away. Examples of such materials, super-invar [19], Zerodur and Cervit [20], are shown in figure **3.6**.

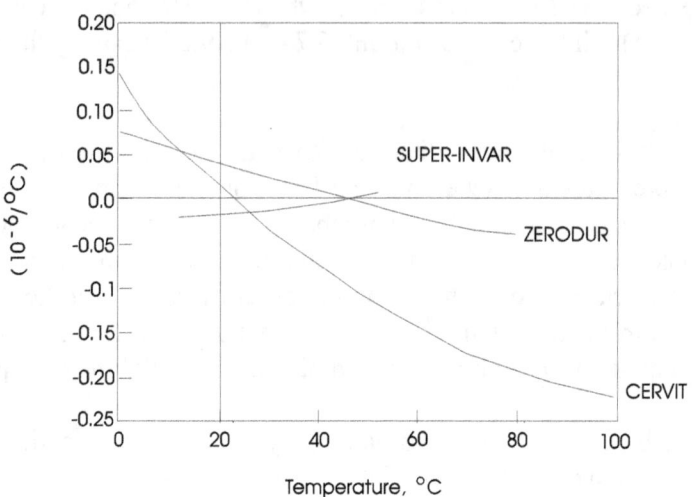

Figure 3.6.   Variation of the thermal expansion coefficient for selected low expansion materials with temperature.

The thermal expansion coefficients, at 20 °C, of commonly used materials in dimensional metrology are shown in table **3.2**.

Table 3.2

Material	Thermal Expansion Coefficient $(10^{-6}/°C)$
Aluminum	24
Free Cutting Brass	20.5
Steel Gauge Block (<25mm)	11.5
Steel Gauge Block (500 mm)	10.6
Ceramic Gauge Block (zirconia)	9.2
Chrome Carbide	8.4
Granite	6.3
Oak (across grain)	5.4
Oak (along grain)	4.9
Tungsten Carbide	4.5
Invar	1.2
Fused Silica	0.55
Zerodur	0.05

To give a more intuitive feel for these numbers, figure 3.7 shows a bar graph of the relative changes in length of 100 mm samples of various materials when taken from 20 °C to 25 °C (68 °F to 77 °C).

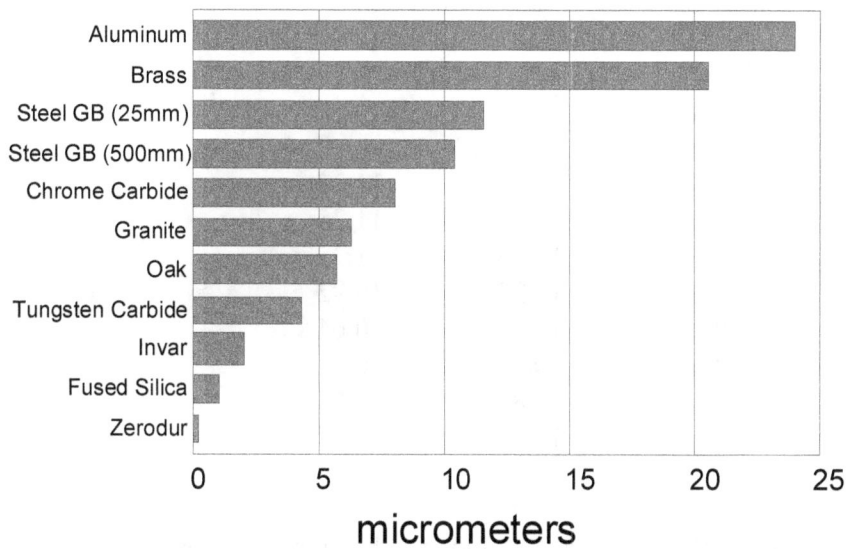

Figure 3.7 Thermal Expansion of 100 mm blocks of various materials from 20 °C to 25 °C

3.3.2 Thermal Expansion Uncertainty

There are two sources of uncertainty in measurements due to the thermal expansion of gauges. These are apparent in thermal expansion equation, 3.2, where we can see that the length of the block depends on our knowledge of both the temperature **and** the thermal expansion coefficient of the gauge. For most measurement systems the uncertainty in the thermometer calibrations is known, either from the manufacturers specifications or the known variations from previous calibrations. For simple industrial thermocouple or thermistor based systems this uncertainty is a few tenths of a degree. For gauging at the sub-micrometer level this is generally insufficient and more sophisticated thermometry is needed.

The uncertainty in the expansion coefficient of the gauge or workpiece is more difficult to estimate. Most steel gauge blocks under 100 mm are within a five tenths of $11.5 \times 10^{-6}/°C$, although there is some variation from manufacturer to manufacturer, and even from batch to batch from the same manufacturer. For long blocks, over 100 mm, the situation is more complicated. Steel gauge blocks have the gauging surfaces hardened during manufacturing so that the surfaces can be properly lapped. This hardening process affects only the 30 to 60 mm of the block near the surfaces. For blocks under 100 mm this is the entire block, and there is no problem. For longer blocks, there is a variable amount of the block in the center which is partially hardened or unhardened. Hardened steel has a higher thermal expansion coefficient than unhardened steel, which means that the longer the block the greater is the unhardened portion and the lower is the coefficient. The measured expansion coefficients of the NIST long gauge blocks are shown in table **3.3**.

Table 3.3. Thermal Expansion Coefficients of NIST Master Steel Gauge Blocks
($10^{-6}/°C$)

Size (in.)	Set 1	Set 2
5	11.41	11.27
6	11.33	11.25
7	11.06	11.35
8	11.22	10.92
10	10.84	10.64
12	10.71	10.64
16	10.80	10.58
20	10.64	10.77

Table **3.3** shows that as the blocks get longer, the thermal expansion coefficient becomes systematically smaller. It also shows that the differences between blocks of the same size can be as large as a few percent. Because of these variations, it is important to use long length standards as near to 20 °C as possible to eliminate uncertainties due to the variation in the expansion coefficient.

As an example, suppose we have a 500 mm gauge block, a thermometer with an uncertainty of 0.1°C, and the thermal expansion coefficient is known to $\pm 0.3 \times 10^{-6}$. The uncertainties when the thermometer reads 20 and 25 degrees are

$$\Delta L = \alpha_L L \times \delta(\Delta T) + \delta(\alpha_L) L \times \Delta T \tag{3.4}$$

where ΔT is the temperature difference (T-20), and $\delta()$ denotes the uncertainty of the quantity within the parentheses.

At 25 °C:

$$\Delta L = (11.5 \times 10^{-6}) \times 500 \times 0.1 + 0.3 \times 10^{-6} \times 500 \times 5 \tag{3.5}$$
$$= 0.58 \, \mu m + .75 \, \mu m$$
$$= 1.33 \, \mu m$$

At 20 °C: when the thermometer reads 20 °C, the worst case error is 0.1 °C

$$\Delta L = (11.5 \times 10^{-6}) \times 500 \times 0.1 + 0.3 \times 10^{-6} \times 500 \times .1 \tag{3.6}$$
$$= 0.58 \, \mu m + 0.02 \, \mu m$$
$$= 0.60 \, \mu m$$

This points out the general need to keep dimensional metrology labs at, or very near 20 °C.

3.4 Elastic Properties

When a force is exerted on any material, the material deforms. For steel and other gauge block materials this effect is small, but not completely negligible. There are two-dimensional effects due to the elastic properties of gauge blocks. The first, and least important, is the compression of blocks under their own weight. When a block is supported horizontally, the force on each point is the weight of the steel above it, and the steel is slightly compressed. The compression is, however, not in the direction of the gauging dimension of the block and the effect is negligible. If the block is set upright, the force is now in the direction of the gauging surfaces, and for very long blocks the weight of the block can become significant. Solved analytically, the change in length of a block is found to be

$$\Delta L = \rho g L^2 / 2E \tag{3.7}$$

Where

ΔL = length of shortening
ρ = density of material
g = acceleration of gravity
L = total length of block
E = Young's modulus for material

For steel gauge blocks, the shrinkage is

$$\Delta L = (7.8 \times 10^3 (kg/m^3) \times 9.8 \, m/s^2 \times L^2)/(2 \times 210 \times 10^9 \, N/m^2) \tag{3.8}$$

$$= 0.18 \times 10^{-6} \times L^2 \text{ in meters.}$$

For a 500 mm gauge block the correction is 45 nm (1.8 µin). The corrections from this formula are made at NIST on demand, but are negligible for blocks less than 300 mm (12 in).

When using long gauge blocks supported horizontally, some care is needed to assure that the block bends properly. Since the sides of the gauge block are not precision surfaces, no matter how flat the surface where it is placed it will touch only at a few points, therefore bending, and in general producing some small angle between the two gauging faces. The proper way to support the block so that the two end faces are parallel, and thereby produce an unambiguous length, is shown in figure 3.8. This assumes, however, that the gauging faces are parallel when the block is vertical.

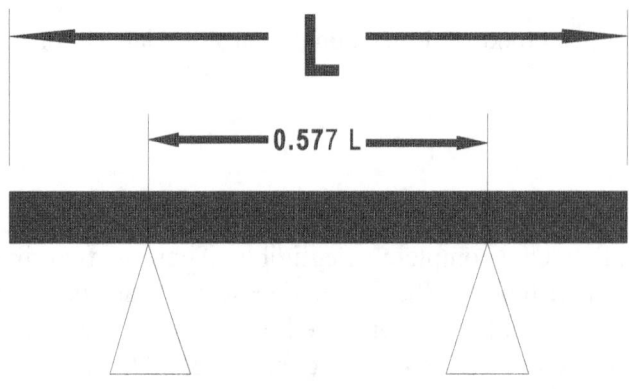

Figure 3.8. Long gauge block supported at its Airy points.

When a block of length L is supported at two positions, 0.577L apart, the end faces will be parallel. These positions are called the Airy points.

3.4.1 Contact Deformation in Mechanical Comparisons

Nearly all gauge block length comparisons or length measurements of objects with gauge blocks are made with contact type comparators where a probe tip contacts a surface under an applied force. Contact between a spherical tip and a plane surface results in local deformation of small but significant magnitude. If the gauge blocks or objects being compared are made of the same material, the measured length difference between them will be correct, since the deformation in each case will be the same. If the materials are different, the length difference will be incorrect by the difference in the amount of deformation for the materials. In such cases, a deformation correction may be applied if its magnitude is significant to the measurement.

Total deformation (probe plus object) is a function of the geometry and elastic properties of the two contacting surfaces, and contact force. Hertz [21] developed formulas for total uniaxial deformation based on the theory of elasticity and by assuming that the bodies are isotropic, that there is no tangential force at contact, and that the elastic limit is not exceeded in the contact area. Many experimenters [22, 23] have verified the reliability of the Hertzian formulas. The formulas given below are from a CSIRO (Australian metrology laboratory) publication that contains formulas for a number of combinations of geometric contact between planes, cylinders and spheres [24]. The gauge block deformations have been tested against other calculations and agree to a few nanometers.

For a spherical probe tip and a flat object surface the uniaxial deformation of the probe and surface together is given by:

(3.9)
$$\alpha = \frac{(3\pi)^{2/3}}{2} \cdot P^{2/3} \cdot (V_1 + V_2)^{2/3} \cdot \left(\frac{1}{D}\right)^{1/3}$$

Where

$V_1 = (1 - \sigma_1^2)/\pi E_1$
σ_1 = Poisson ratio of sphere
E_1 = elastic modulus of sphere

$V_2 = (1 - \sigma_2^2)/\pi E_2$
σ_2 = Poisson ratio of block
E_2 = elastic modulus of block

P = force
D = diameter of sphere

Some example deformations for a 6 mm diameter diamond stylus, and common gauge block materials, at various pressures are given in table 3.4.

Table 3.4 Deformations at interface in micrometers (µin in parenthesis)

Material	Force in Newtons		
	0.25	0.50	0.75
Fused Silica	0.13 (5.2)	0.21 (8.3)	0.28 (11.2)
Steel	0.07 (2.7)	0.11 (4.4)	0.14 (5.7)
Chrome Carbide	0.06 (2.2)	0.12 (3.4)	0.12 (4.6)
Tungsten Carbide	0.04 (1.6)	0.06 (2.5)	0.08 (3.2)

The gauge block comparators at NIST use 6 mm diameter tips, with forces of 0.25 N on the bottom probe and 0.75 N on the top probe. The trade-offs involved in tip radius selection are:

1. The larger the probe radius the smaller the penetration, thus the correction will be smaller and less dependent on the exact geometry of the tip radius.

2. The smaller the tip radius the greater its ability to push foreign matter, such as traces of oil, water vapor, or dust out of the way when the block is dragged between the tips.

There are comparator tips used by some laboratories that have radii as large as 10 to 20 mm. These large radius tips have small penetrations, although generally more than 20 nm, and thus some correction still must be made.

3.4.2 Measurement of Probe Force and Tip Radius

Reliability of computed deformation values depends on careful measurement of probe force and, especially, of probe tip radius. Probe force is easily measured with a force gauge (or a double pan balance and a set of weights) reading the force when the probe indicator meter is at mid-scale on the highest magnification range.

Probe tip inspection and radius measurement are critical. If the tip geometry is flawed in any way it will not follow the Hertz predictions. Tips having cracks, flat spots, chips, or ellipticity should be replaced and regular tip inspection must be made to insure reliability.

An interference microscope employing multiple beam interferometry is the inspection and measurement method used at NIST [25]. In this instrument an optical flat is brought close to the tip, normal to the probe axis, and monochromatic light produces a Newton Ring fringe pattern which is magnified though the microscope lens system. The multiple beam aspect of this instrument is produced by special optical components and results in very sharp interference fringes which reveal fine details in the topography of the tip.

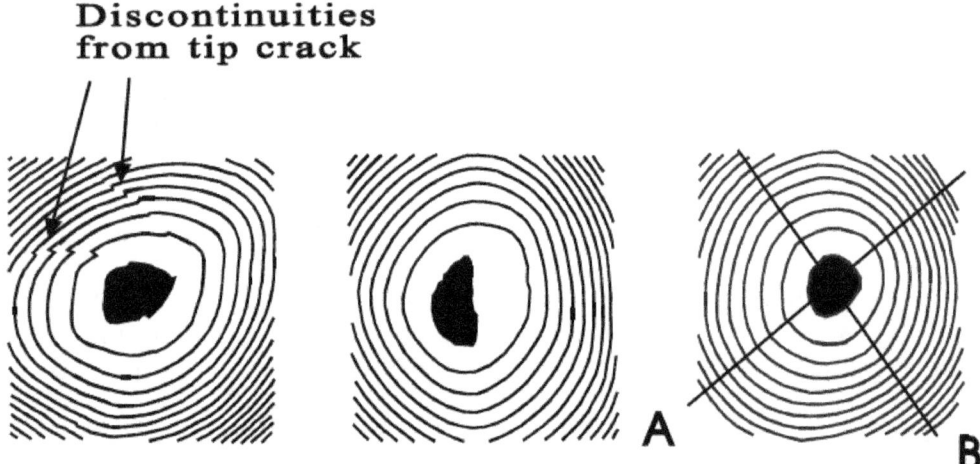

Figure 3.9 a, b, and c. Examples of microinterferograms of diamond stylus tips.

Figure **3.9** are multiple beam interference micrographs of diamond probe tips. The pictures have been skeleltonized so that the important features are clear. The micrograph is a "contour map" of the tip so that all points on a given ring are equidistant from the reference optical flat. This can be expressed mathematically as

$$N\lambda = 2t \qquad (3.10)$$

where N is the number (order) of the fringe, counting from the center, λ is the wavelength of the light and t is the distance of the ring from the optical flat. The zero order fringe at the center is where the probe tip is in light contact with the optical flat and t is nearly zero. This relationship is used to calculate the tip radius.

Tip condition is readily observed from the micrograph. In figure **3.9a** a crack is seen in the tip as well as some ellipticity. The crack produces sharp breaks and lateral displacement of the fringes along the crack. Figure **3.9b** shows a sharp edge at the tip center. An acceptable tip is shown in figure **3.9c**. It is very difficult to produce a perfectly spherical surface on diamond because the diamond hardness is not isotropic, i.e., there are "hard" directions and "soft" directions, and material tends to be lapped preferentially from the "soft" directions.

Radius measurement of a good tip is relatively simple. Diameters of the first five rings are measured from the photograph along axes A and B, as in figure **3.9c**. If there is a small amount of ellipticity, A and B are selected as the major and minor axes. Then

$$r_d = \frac{d_n}{2M} \qquad (3.11)$$

where r_d is the actual radius of the nth Newton ring, d_n is the ring average diameter (of A or B) measured on the micrograph, and M is the microscope magnification. Substituting the ring diameter measurements in the equation will result in 5 radii, r_1 through r_5 and from these a radius of curvature between consecutive rings is calculated:

$$R_i = \frac{r_{i+1}^2 - r_i^2}{\lambda} \qquad (3.12)$$

for i = 1 to 4. The average of these four values is used as the tip radius.

The preceding measured and calculated values will also serve to evaluate tip sphericity. If the average difference between the five A and B ring diameters exceeds 10 percent of the average ring diameter, there is significant lack of sphericity in the tip. Also, if the total spread among the four tip radius values exceeds 10 percent of the average R there is significant lack of sphericity. These tests check sphericity around two axes so it is important that a tip meet both requirements or it will not follow the Hertz prediction.

Our current laboratory practice uses only like materials as master blocks for comparisons. By having one set of steel masters and one set of chrome carbide masters, the only blocks which have deformation corrections are tungsten carbide. We have too small a customer base in this material to justify the expense of a third master set. This practice makes the shape of the comparator tips unimportant, except for cracks or other abnormalities which would scratch the blocks.

3.5 Stability

No material is completely stable. Due to processes at the atomic level all materials tend to shrink or grow with time. The size and direction of dimensional change are dependent on the fabrication processes, both the bulk material processing as well as the finishing processing. During the 1950's the gauge block manufacturers and an interdisciplinary group of metallurgists, metrologists and statisticians from NIST (NBS at the time) did extensive studies of the properties of steel gauge blocks to optimize their dimensional stability. Blocks made since that era are remarkably stable compared to their predecessors, but not perfect. Nearly all NIST master gauge blocks are very stable. Two typical examples are shown in figure 3.10. Because most blocks are so stable, we demand a measurement history of at least 5 years before accepting a non-zero slope as real.

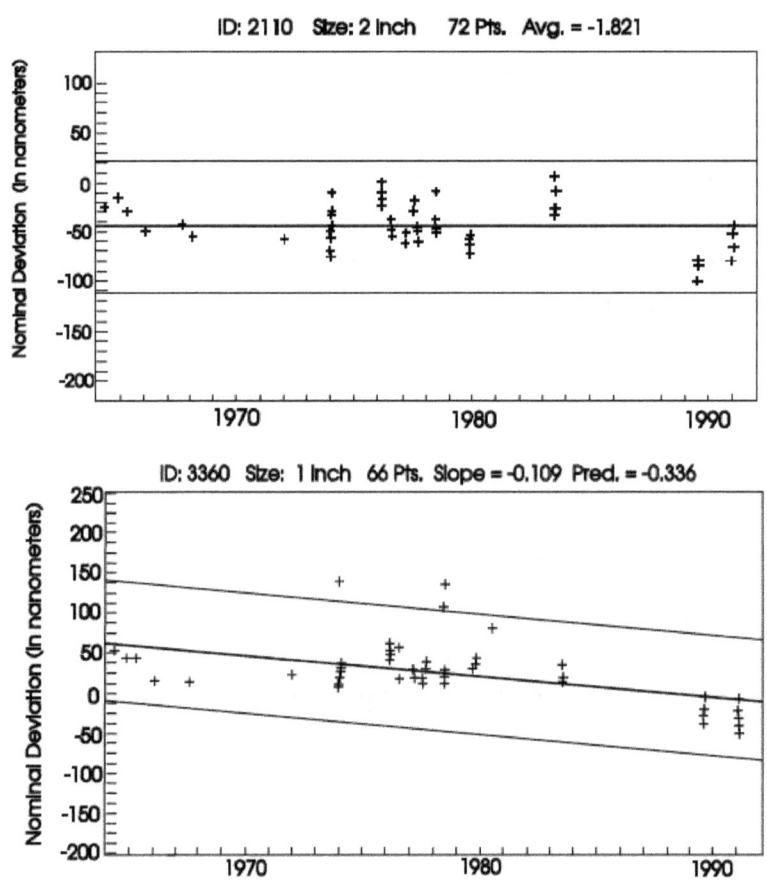

Figure 3.10. Examples of the dimensional stability of NIST master gauge blocks.

A histogram of the growth rates of our master blocks, which have over 15 years of measurement history, is shown in figure **3.11**. Note that the rate of change/unit of length is the pertinent parameter, because most materials exhibit a constant growth rate per unit of length.

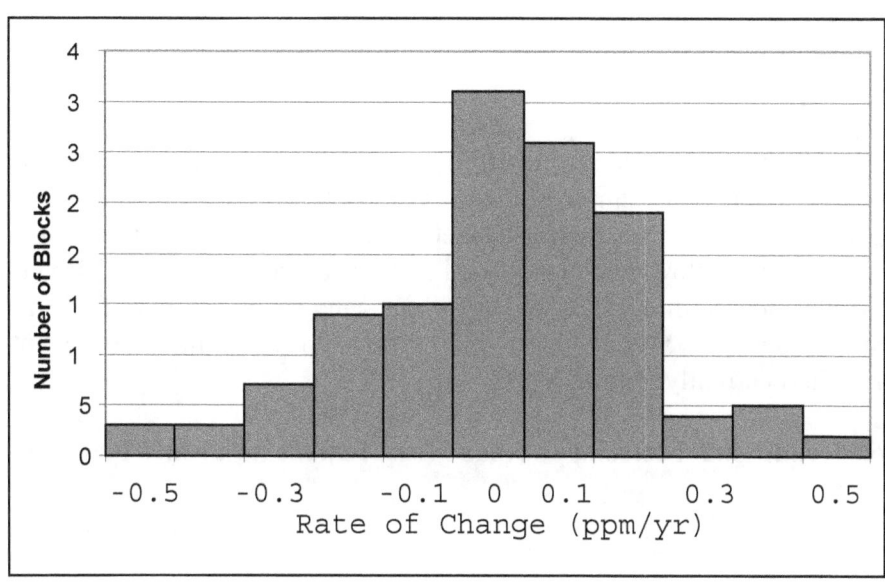

Figure 3.11. Histogram of the growth rates of NIST master blocks.

There have been studies of dimensional stability for other materials, and a table of typical values [26-31] is given in table **3.5**. In general, alloys and glassy materials are less stable than composite materials. For the average user of gauge blocks the dimensional changes in the time between calibration is negligible.

Table 3.5

Material	Stability (1 part in 10^6/yr)
Zerodur	-0.7 to -0.03
Corning 7971 ULE	-0.14, 0.07, 0.06
Corning 7940 (fused Silica)	-0.18, -0.18
Cer-Vit C-1010	0.54, -0.18
LR-35 Invar	2.05
Super Invar	0.0
52100 Gauge Block Steel	0.01
Brass	-1.2

Steel, chrome carbide and tungsten carbide have all proven to be very stable materials for gauge blocks. The newer ceramic materials may also be stable, but no studies have been made of either the material or gauge blocks made from it. Until studies are available, a shorter calibration cycle for the first few years should be considered to assure that the blocks are suitably stable.

4. Measurement Assurance Program

4.1 Introduction

One of the primary problems in metrology is to estimate the uncertainty of a measurement. Traditional methods, beginning with simple estimates by the metrologist based on experience gradually developed into the more formal error budget method. The error budget method, while adequate for many purposes, is slowly being replaced by more objective methods based on statistical process control ideas. In this chapter we will trace the development of these methods and derive both an error budget for gauge block calibrations and present the measurement assurance (MAP) method currently used at NIST.

4.2 A comparison: Traditional Metrology versus Measurement Assurance Programs

4.2.1 Tradition

Each measurement, in traditional metrology, was in essence a "work of art." The result was accepted mainly on the basis of the method used and the reputation of the person making the measurement. This view is still prevalent in calibrations, although the trend to statistical process control in industry and the evolution of standards towards demanding supportable uncertainty statements will eventually end the practice.

There are circumstances for which the "work of art" paradigm is appropriate or even inevitable as, for example, in the sciences where most experiments are seldom repeated because of the time or cost involved, or for one of a kind calibrations . Since there is no repetition with which to estimate the repeatability of the measurement the scientist or metrologist is reduced to making a list of possible error sources and estimating the magnitude of these errors. This list, called the error budget, is often the only way to derive an uncertainty for an experiment.

Making an accurate error budget is a difficult and time consuming task. As a simple example, one facet of the error budget is the repeatability of a gauge block comparison. A simple estimation method is to measure the same block repeatedly 20 or more times. The standard deviation from the mean then might be taken as the short term repeatability of the measurement. To get a more realistic estimate that takes into consideration the effects of the operator and equipment, the metrologist makes a GR&R (gauge repeatability and reproducibility) study. In this test several blocks of different length are calibrated using the normal calibration procedure. The calibrations are repeated a number of times by different operators using different comparators. This data is then analyzed to produce a measure of the variability for the measurement process. Even a study as large as this will not detect the effects of long term variability from sources such as drift in the comparator or thermometer calibrations. Unfortunately, these auxiliary experiments are seldom done in practice and estimates based on the experimenter's experience are substituted for experimentally verified values. These estimates often reflect the experimenter's optimism more than reality.

The dividing line between the era of traditional metrology and the era of measurement assurance programs is clearly associated with the development of computers. Even with simple comparison procedures, the traditionalist had to make long, detailed hand computations. A large percentage of calibration time was spent on mathematical procedures, checking and double checking hand computations. The calibration of a meter bar could generate over a hundred pages of data and calculations. Theoretical work revolved around finding more efficient methods, efficient in the sense of fewer measurements and simpler calculations.

4.2.2 Process Control: A Paradigm Shift

Measurement assurance has now become an accepted part of the national standards system. According to the ANSI/ASQC Standard M-1, "American National Standard for Calibration Systems" the definition of measurement assurance is as follows:

> 2.12 Measurement Assurance Method
>
> A method to determine calibration or measurement uncertainty based on systematic observations of achieved results. Calibration uncertainty limits resulting from the application of Measurement Assurance Methods are considered estimates based upon objective verification of uncertainty components.

Admittedly, this definition does not convey much information to readers who do not already understand measurement. To explain what the definition means, we will examine the more basic concept of process control and show how it relates to measurement and calibration. Measurement assurance methods can be thought of as an extension of process control.

The following is an example of control of a process using a model from the chemical industry. In a chemical plant the raw materials are put into what is essentially a black box. At the molecular level, where reactions occur, the action is on a scale so small and so fast that no human intervention is possible. What can be measured are various bulk properties of the chemicals; the temperature, pressure, fluid flow rate, etc. The basic model of process control is that once the factory output is suitable, if the measurable quantities (process control variables) are kept constant the output will remain constant. Thus the reactions, which are not observable, are monitored and characterized by these observable control variables.

Choosing control variables is the most important part of the control scheme. A thermometer placed on a pipeline or vat may or may not be a useful measure of the process inside. Some chemical reactions occur over a very large range of temperatures and the measurement and control of the temperature might have virtually no effect on the process. In another part of the plant the temperature may be vitally important. The art and science of choosing the process control measurements is a large part of the chemical engineering profession.

Defining the measurement process as a production process evolved in the early 1960's [33], a time coincident with the introduction of general purpose computers on a commercial basis. Both the philosophy and scope of measurement assurance programs are a direct result of being able to store and recall large amounts of data, and to analyze and format the results in many different ways. In this model, the calibration procedure is a process with the numbers as the process output. In our case, the length of a gauge block is the output.

The idea of a measurement assurance program is to control and statistically characterize a measurement process so that a rational judgement about the accuracy the process can be made. Once we accept the model of measurement as a production process we must examine the process for possible control variables, i.e., measurable quantities other than the gauge block length which characterize the process.

For example, suppose we compare one gauge block with one "master" block. If one comparison is made, the process will establish a number for the length of the block. An obvious extension of this process is to compare the blocks twice, the second measurement providing a measure of the repeatability of the process. Averaging the two measurements also will give a more accurate answer for two reasons. First, statistics assures us that the mean of several measurements has a higher probability of being the accurate than any single measurement. Secondly, if the two measurements differ greatly we can repeat the measurement and discard the outlying measurement as containing a blunder. Thus the repeatability can be used as a process control variable.

The next extension of the method is to make more measurements. It takes at least four measurements for statistical measures of variability to make much sense. At this point the standard deviation becomes a possible statistical process control (SPC) parameter. The standard deviation of each measurement is a measure of the short term repeatability of the process which can be recorded and compared to the repeatability of previous measurements. When this short term repeatability is much higher than its historical value we can suspect that there is a problem with the measurement, i.e., the process is out of control. Since the test is statistical in nature we can assign a confidence level to the decision that the process is in or out of control. An analysis of the economic consequences of accepting bad or rejecting good calibrations can be used to make rational decisions about the control limits and confidence levels appropriate to the decisions.

Another extension is to have two master blocks and compare the unknown to both of them. Such a procedure can also give information about the process, namely the observed difference between the two masters. As in the previous example, this difference can be recorded and compared to the differences from previous calibrations. Unlike the previous case this process control variable measures long term changes in the process. If one of the blocks is unstable and grows, over a year the difference between the masters will change and an examination of this control parameter will detect it. Short term process repeatability is not affected by the length changes of the blocks because the blocks do not change by detectable amounts during the few minutes of a comparison.

Thus, comparing a gauge block with two masters according to a measurement design provides, not only a value for the block, but also an estimate of short term process repeatability, and in time, an estimate of long term process variability and a check on the constancy of the masters. All of this is inexpensive only if the storage and manipulation of data is easy to perform and fast, tasks for which computers are ideal.

The complexity of a measurement assurance program depends upon the purpose a particular measurement is to serve. NIST calibration services generally aim at providing the highest practical level of accuracy. At lower accuracy levels the developer of a measurement system must make choices, balancing the level of accuracy needed with costs. A number of different approaches to measurement assurance system design are discussed by Croarkin [34].

4.2.3 Measurement Assurance: Building a Measurement Process Model

Assigning a length value to a gauge block and determining the uncertainty of that value is not a simple task. As discussed above, the short and long term variability of the measurement process are easily measured by redundant measurements and a control measurement (the difference between two masters) in each calibration. This is not enough to assure measurement accuracy. SPC only attempts to guarantee that the process today is the same as the process in the past. If the process begins flawed, i.e., gives the wrong answer, process control will only guarantee that the process continues to give wrong answers. Obviously we need to ascertain the sources of error other than the variability measured by the SPC scheme.

Many auxiliary parameters are used as corrections to the output of the measurement procedure. These corrections must be measured separately to provide a complete physical model for the measurement procedure. Examples of these parameters are the phase change of light on reflection from the gauge block during interferometric measurements, the penetration depth of the gauge block stylus into the block during mechanical comparisons, and thermal expansion coefficients of the blocks during all measurements. These auxiliary parameters are sources of systematic error.

To assess systematic errors, studies must be made of the effects of factors not subject to control, such as the uncertainty of thermometer and barometer calibrations, or variations in the deformability of gauge blocks. Combining these systematic errors with the known random error, one arrives at a realistic statement of the accuracy of the calibration.

In summary, the MAP approach enables us to clearly establish limitations for a particular measurement method. It combines process control with detailed process modeling. The process control component provides a means to monitor various measurement parameters throughout the system and provides estimates of the random error. The calibration model allows us to obtain reasonable estimates of the systematic uncertainties that are not sampled by the statistics derived from an analysis of the process control parameters. This leads to a detailed understanding of the measurement system and an objective estimate of the measurement uncertainty. It also creates a reliable road map for making process changes that will increase calibration accuracy.

4.3 Determining Uncertainty

4.3.1 Stability

All length measurement processes are, directly or indirectly, comparative operations. Even the simplest concept of such a process assumes that the basic measurement unit (in our case the meter) is constant and the object, procedures, equipment, etc., are stable.

As an example, before 1959 the English inch was defined in terms of an official yard bar and the American inch was defined in terms of the meter. By 1959 the English inch had shrunk compared to the American inch because the yard bar was not stable and was shrinking with respect to the meter. This problem was solved by adopting the meter as the primary unit of length and defining the international inch as exactly 25.4 mm.

Similarly, the property to be measured must be predictable. If a measurement process detects a difference between two things, it is expected that repeated measures of that difference should agree reasonably well. In the absence of severe external influence, one does not expect things to change rapidly.

There is a difference between stability and predictability as used above. Repeated measurements over time can exhibit a random-like variability about a constant value, or about a time dependent value. In either case, if the results are not erratic (with no unexpected large changes), the process is considered to be predictable. Gauge blocks that are changing length at a constant rate can be used because they have a predictable length at any given time. Stability means that the coefficients of time dependent terms are essentially zero. Stability is desirable for certain uses, but it is not a necessary restriction on the ability to make good measurements.

4.3.2 Uncertainty

A measurement process is continually affected by perturbations from a variety of sources. The random-like variability of repeated measurements is a result of these perturbations. Random variability implies a probability distribution with a range of variability that is not likely to exceed test limits. Generally a normal distribution is assumed.

Traditionally the second type of error is called systematic error, and includes uncertainties that come from constants that are in error and discrepancies in operational techniques. The systematic error, expressed as a single number, is an estimate of the offset of the measurement result from the true value. For example, for secondary laboratories using NIST values for their master blocks the uncertainty reported by NIST is a systematic error. No matter how often the master blocks are used the offset between the NIST values and the true lengths of the blocks remains the same. The random error and systematic error are combined to determine the uncertainty of a calibration.

A new internationally accepted method for classifying and combining errors has been developed by the International Bureau of Weights and Measures (BIPM) and the International Organization for Standardization (ISO). This method seems more complicated than the traditional method, and is presented in detail in the ISO "Guide to the Expression of Uncertainty in Measurement" [35]. A short summary of the method and how it will be applied at NIST is given in reference [36]. Quoting ISO, the core concepts are:

> 1. The uncertainty in the result of a measurement generally consists of several components which may be grouped into two categories according to the way in which their numerical value is estimated:
>
>> A - those which are evaluated by statistical methods,
>> B - those which are evaluated by other means.
>
> There is not always a simple correspondence between the classification into categories A or B and the previously used classification into "random" and "systematic" uncertainties. The term "systematic uncertainty" can be misleading and should be avoided.
>
> Any detailed report of the uncertainty should consist of a complete list of the components, specifying for each the method used to obtain its numerical value.
>
> 2. The components in category A are characterized by estimated variances, s_i^2, (or the estimated "standard deviations" s_i) and the number of degrees of freedom, v_i. Where appropriate, the estimated covariances should be given.
>
> 3. The components in category B should be characterized by quantities u_j^2, which may be considered as approximations to the corresponding variances, the existence of which is assumed. The quantities u_j^2 may be treated like variances and the quantities u_j like standard deviations. Where appropriate, the covariances should be treated in a similar way.
>
> 4. The combined uncertainty should be characterized by the numerical value obtained by applying the usual method for the combination of variances. The combined uncertainty and its components should be expressed in the form of "standard deviations."
>
> 5. If, for particular applications, it is necessary to multiply the combined uncertainty by a factor to obtain an overall uncertainty, the multiplying factor used must always be stated.

The random errors, being determined statistically, are type A errors. The errors that made up the systematic errors become the sources of type B uncertainty. The main difference is the method for estimating the type B uncertainty. The traditional systematic error was estimated by the metrologist with little or no guidance. The ISO guide gives specific criteria for the estimation of type B uncertainty. NIST has adopted the new international method for reporting calibration uncertainties. In addition to adopting the new method, NIST has adopted k=2 as the coverage factor, rather than the more traditional 3σ, to align its reported uncertainty with the current International practice.

4.3.3 Random Error

There are very few sources of truly random error. Some examples are 1/f and shot noise in electronic components. These contributions to the uncertainty of most dimensional measurements are very small. Most of the contributions to the random errors are systematic effects from sources that are not under experimental control. Examples are floor and air vibrations, thermal drift, and operator variability. Many of these error sources are too costly to control.

These effects can be grouped into two categories: long term effects in which the effect does not significantly change in the time required for a given sequence of measurements and short term effects which vary significantly in the course of a single measurement or measurements made over a short time interval. Typical long term effects are changes in length of the master blocks or slow changes in the comparator electronic magnification.

An example of smooth continuous short term effects are thermal drift of the environment during the measurement. Another category of short term effects are those which are instantaneous, or step-like, in nature. A simple example is a dust particle left on a gauge block. Depending on the exact point of contact the gauge length may be shifted by the size of the particle. In many cases, "shocks" or vibrations can cause temporary shifts in the instrument output.

These short term variations make up the within-group variability usually expressed as a standard deviation, σ_w. The within-group variability of the measurement process is the most familiar process parameter, and is generally called repeatability. It is easily demonstrated in a repeated sequence of measurements of the same thing in a short time interval. Practically all measurements should be repeated several times. Within-group variability is generally limited by the degree that certain types of perturbations are controlled and by factors such as operator skills, instrument quality, and attention to detailed procedure. Sources contributing to within variability are not easily identified and process improvement in terms of reducing σ_w is obtained more frequently by trial and error than by design.

Process accuracy is often judged on the basis of within-group variability. Such a judgment, however, is usually erroneous. A more reliable assessment is the total variability that is the variability of a long sequence of measurements. Repeating a given measurement over a time interval sufficiently long to reflect the influence of all possible perturbations establishes a total process standard deviation, σ_t.

The primary way to improve total variability is to modify the process by reducing the variability associated with all identifiable perturbations. This can be very tedious and expensive, but is necessary to produce the highest quality measurements.

Another straightforward way to reduce variability is to place severe restrictions on the characteristics of objects being measured and on the measurement environment. Reducing the complexity of the object reduces the possible error sources. Thus traditional metrology focused on simple geometric forms, spheres, cylinders and blocks as the mainstays of length standards.

By making all measurements at or near standard conditions, 20 °C for example, many other errors are reduced. Restricting the number and severity of possible errors brings the uncertainty estimate closer to the truth.

4.3.4 Systematic Error and Type B Uncertainty

The sources of systematic error and type B uncertainty are the same. The only major difference is in how the uncertainties are estimated. The estimates for systematic error which have been part of the NIST error budget for gauge blocks have been traditionally obtained by methods which are identical or closely related to the methods recommended for type B uncertainties.

There are several classes of type B uncertainty. Perhaps the most familiar is instrument scale zero error (offset). Offset systematic errors are not present in comparative measurements provided that the instrument response is reasonably linear over the range of difference that must be measured. A second class of type B uncertainty comes from supplemental measurements. Barometric pressure, temperature, and humidity are examples of interferometric supplemental measurements used to calculate the refractive index of air, and subsequently the wavelength of light at ambient conditions. Other examples are the phase shift of light on reflection from the block and platen, and the stylus penetration in mechanical comparisons. Each supplemental measurement is a separate process with both random variability and systematic effects. The random variability of the supplemental measurements is reflected in the total process variability. Type B Uncertainty associated with supplemental data must be carefully documented and evaluated.

One practical action is to "randomize" the type B uncertainty by using different instruments, operators, environments and other factors. Thus, variation from these sources becomes part of the type A uncertainty. This is done when we derive the length of master blocks from their interferometric history. At NIST, we measure blocks using both the Hilger[1] and Zeiss[1] interferometers and recalibrate the barometer, thermometers, and humidity meters before starting a round of calibrations. As a measurement history is produced, the effects of these factors are randomized. When a block is replaced this "randomization" of various effects is lost and can be regained only after a number of calibrations.

Generally, replacement blocks are not put into service until they have been calibrated interferometrically a number of times over three or more years. Since one calibration consists of 3 or more wrings, the minimum history consists of nine or more measurements using three different calibrations of the environmental sensors. While this is merely a rule of thumb, examination of the first 10 calibrations of blocks with long histories appear to assume values and variances which are consistent with their long term values.

The traditional procedure for blocks that do not have long calibration histories is to evaluate the type B uncertainty by direct experiment and make appropriate corrections to the data. In many

---

[1] Certain commercial instruments are identified in this paper in order to adequately specify our measurement process. Such identification does not imply recommendation or endorsement by NIST.

cases the corrections are small and the uncertainty of each correction is much smaller than the process standard deviation. For example, in correcting for thermal expansion, the error introduced by a temperature error of 0.1 °C for a steel gauge block is about 1 part in 10^6. This is a significant error for most calibration labs and a correction must be applied. The uncertainty in this correction is due to the thermometer calibration and the uncertainty in the thermal expansion coefficient of the steel. If the measurements are made near 20 °C the uncertainty in the thermal coefficient (about 10%) leads to an uncertainty in the correction of 1 part in 10^7 or less; a negligible uncertainty for many labs.

Unfortunately, in many cases the auxiliary experiments are difficult and time consuming. As an example, the phase correction applied to interferometric measurements is generally about 25 nm (1 μin). The 2σ random error associated with measuring the phase of a block is about 50 nm (2μin). The standard deviation of the mean of n measurements is $n^{-1/2}$ times the standard deviation of the original measurements. A large number of measurements are needed to determine an effect which is very small compared to the standard deviation of the measurement. For the uncertainty to be 10% of the correction, as in the thermal expansion coefficient example, n must satisfy the relation:

$$10\% \text{ of } 25 < \frac{50}{\sqrt{n}} \quad \text{(in nanometers)} \tag{4.1}$$

Since this leads to n=400, it is obvious this measurement cannot be made on every customer block. Phase measurements made for a random sample of blocks from each manufacturer have shown that the phase is nearly the same for blocks made (1) of the same material, (2) by the same lapping procedure and (3) by the same manufacturer. At NIST, we have made supplementary measurements under varying conditions and calculated compensation factors for each manufacturer (see appendix C). A corollary is that if you measure your master blocks by interferometry, it helps to use blocks from only one or two manufacturers.

Our next concern is whether the corrections derived from these supplementary measurements are adequate. A collection of values from repeated measurements should be tested for correlation with each of the supplementary measurements, and their various combinations. If correlation is indicated then the corrections that are being applied are not adequate. For example, if a correlation exists between the length of gauge blocks previously corrected for temperature and the observed temperature of the block, then either the temperature is being measured at an inappropriate location or the correction scheme does not describe the effect that is actually occurring. Corrective action is necessary.

Low correlation does not necessarily indicate that no residual type B uncertainty are present, but only that the combined type B uncertainty are not large relative to the total standard deviation of the process. There may be long term systematic error effects from sources not associated with the current supplemental measurements. It is relatively easy to demonstrate the presence or absence of such effects, but it may be difficult to reduce them. If both the short and long term standard deviations of the process are available, they are expected to be nearly equal if there are no unresolved sources of variation. Frequently such is not the case. In the cases where the long term SD is distinctly larger than the short term SD there are uncorrected error sources at work.

The use of comparison designs, described in chapter 5, facilitates this type of analysis. The within group variability, σ_w, is computed for each measurement sequence. Each measurement sequence includes a check standard that is measured in each calibration. The total standard deviation, σ_t is computed for the collection of values for the check standard. The inequality $\sigma_t >> \sigma_w$ is taken as evidence of a long term systematic effect, perhaps as yet unidentified. For the NIST system a very large number of measurements have been made over the years and we have found, as shown in table 4.1, that except for long blocks there appear to be no uncorrected long term error sources. The differences for long blocks are probably due to inadequate thermal preparation or operator effects caused by the difficult nature of the measurement.

Table 4.1

Comparison of Short Term and Long Term Standard Deviations
for Mechanical Gauge Block Comparisons (in nanometers)

Group	Range			$\sigma_{\text{short term}}$	$\sigma_{\text{long term}}$
14	1.00 mm	to	1.09 mm	5	6
15	1.10 mm	to	1.29 mm	5	5
16	1.30 mm	to	1.49 mm	5	5
17	1.50 mm	to	2.09 mm	5	5
18	2.10 mm	to	2.29 mm	5	5
19	2.30 mm	to	2.49 mm	5	5
20	2.50 mm	to	10 mm	5	5
21	10.5 mm	to	20 mm	5	7
22	20.5 mm	to	50 mm	6	8
23	60 mm	to	100 mm	8	18
24		125 mm		9	19
		150 mm		11	38
		175 mm		10	22
		200 mm		14	37
		250 mm		11	48
		300 mm		7	64
		400 mm		11	58
		500 mm		9	56

4.3.5 Error Budgets

An error budget is a systematic method for estimating the limit of error or uncertainty for a measurement. In this method, one lists all known sources of error which might affect the measurement result. In the traditional method, a theoretical analysis of error sources is made to provide estimates of the expected variability term by term. These estimates are then combined to obtain an estimate of the total expected error bounds.

The items in an error budget can be classified according to the way they are most likely to affect the uncertainty. One category can contain items related to type B uncertainty; another can relate to σ_w; and a third category can relate to σ_t. Since all sources of short term variability also affect the long term variability, short term effects are sometimes subtracted to give an estimate of the effects of only the sources of long term variability. This long term or "between group variability", σ_B, is simply calculated as $(\sigma_t - \sigma_w)$. The example in table 4.2 shows a tentative disposition for the mechanical comparison of gauge blocks.

Table 4.2

Error Source	Category
Uncertainty of Master Block	systematic
Stylus diameter, pressure and condition	long term
Assigned Temperature Coefficients	systematic
Temperature difference between blocks	long term
Stability of Blocks	long term
Environment: vibration, thermal changes, operator effects	short term
Block flatness/ anvil design	systematic
Block geometry: effect of flatness and parallelism on measurement with gauging point variability	short term
Comparator calibration	long term
Comparator repeatability	short term

Most of the error sources in table 4.2 can be monitored or evaluated by a judicious choice of a comparison design, to be used over a long enough time so that the perturbations can exert their full effect on the measurement process. Two exceptions are the uncertainties associated with the assigned starting values, or restraint values, and the instrument calibration.

Table 4.3 shows error budget items associated with gauge block measurement processes. Listed (1σ level) uncertainties are typical of the NIST measurement process for blocks under 25 mm long. The block nominal length, L, is in meters.

Table 4.3

Error Sources	Resulting Length Uncertainty in μm	Note
Repeatability	0.004 + 0.12 L	(1)
Contact Deformation	0.002	(2)
Value of Master Block	0.012 + 0.08 L	
Temperature Difference Between Blocks	0.17 L	(3)
Uncertainty of Expansion Coefficient	0.20 L	(4)

Note 1. Repeatability is determined by analyzing the long term control data. The number reported is the $1\sigma_t$ value for blocks under 25 mm.

Note 2. For steel and chrome carbide, blocks of the same material are used in the comparator process and the only error is from the variation in the surface hardness from block to block. The number quoted is an estimate. For other materials the estimate will be higher.

Note 3. A maximum temperature difference of 0.030 °C is assumed. Using this as the limit of a rectangular distribution yields a standard uncertainty of 0.17L.

Note 4. A maximum uncertainty of 1×10^{-6}/°C is assumed in the thermal expansion coefficient, and a temperature offset of 0.3 °C is assumed. For steel and chrome carbide blocks we have no major differential expansion error since blocks of the same material are compared. For other materials the estimate will be higher.
These errors can be combined in various ways, as discussed in the next section.

4.3.6 Combining Type A and Type B uncertainties

Type A and type B uncertainties are treated as equal, statistically determined and if uncorrelated are combined in quadrature.

In keeping with current international practice, the coverage factor k is set equal to two (k=2) on NIST gauge block uncertainty statements.

$$U = \text{Expanded Uncertainty} = k\sqrt{\sum A^2 + \sum B^2} \qquad (4.2)$$

Combining the uncertainty components in Table 4.3 according to these rules gives a standard uncertainty (coverage factor k=2)

$$U = \text{Expanded Uncertainty} = 0.026 + 0.32L \qquad (4.3)$$

Generally, secondary laboratories can treat the NIST stated uncertainty as an uncorrelated type B uncertainty. There are certain practices where the NIST uncertainty should be used with caution. For example, if a commercial laboratory has a gauge block measured at NIST every year for nine years, it is good practice for the lab to use the mean of the nine measurements as the length of the block rather than the last reported measurement. In estimating the uncertainty of the mean, however, some caution is needed. The nine NIST measurements contain type B uncertainties that are highly correlated, and therefore cannot be handled in a routine manner. Since the phase assigned to the NIST master blocks, the penetration, wringing film thickness and thermal expansion differences between the NIST master and the customer block do not change over the nine calibrations, the uncertainties due to these factors are correlated. In addition, the length of the NIST master blocks are taken from a fit to all of its measurement history. Since most blocks have over 20 years of measurement history, the new master value obtained when a few new measurements are made every 3 years is highly correlated to the old value since the fit is heavily weighted by the older data. To take the NIST uncertainty and divide it by 3 (square root of 9) assumes that all of the uncertainty sources are uncorrelated, which is not true.

Even if the standard deviation from the mean was calculated from the nine NIST calibrations, the resulting number would not be a complete estimate of the uncertainty since it would not include the effects of the type B uncertainties mentioned above. The NIST staff should be consulted before attempting to estimate the uncertainty in averaged NIST data.

4.3.7 Combining Random and Systematic Errors

There is no universally accepted method of combining these two error estimates. The traditional way to combine random errors is to sum the variances characteristic of each error source. Since the standard deviation is the square root of the variance, the practice is to sum the standard deviations in quadrature. For example, if the measurement has two sources of random error characterized by σ_1 and σ_2, the total random error is

$$\sigma_{tot} = [\sigma_1^2 + \sigma_2^2]^{1/2}. \qquad (4.4)$$

This formula reflects the assumption that the various random error sources are independent. If there are a number of sources of variation and their effects are independent and random, the probability of all of the effects being large and the same sign is very small. Thus the combined error will be less than the sum of the maximum error possible from each source.

Traditionally, systematic errors are not random (they are constant for all calibrations) and may be highly correlated, so nothing can be assumed about their relative sizes or signs. It is logical then to simply add the systematic errors to the total random error:

$$\text{Total Uncertainty (traditions)} = 3\sigma_{tot} + \text{Systematic Error.} \tag{4.5}$$

The use of 3σ as the total random error is fairly standard in the United States. This corresponds to the 99.7% confidence level, i.e., we claim that we are 99.7% confident that the answer is within the 3σ bounds. Many people, particularly in Europe, use the 95% confidence level (2σ) or the 99% confidence level (2.5σ) rather than the 99.7% level (3σ) used above. This is a small matter since any time an uncertainty is stated the confidence level should also be stated. The uncertainty components in Table 4.3 are all systematic except the repeatability, and the uncertainty calculated according to the traditional method is

$$\text{Total Uncertainty (traditional)} = 3(0.004 + 0.12L) + 0.002 + 0.012 + 0.08L + 0.17L + 0.2L$$

$$\text{Total Uncertainty (traditional)} = 0.026 + 0.81L$$

4.4 The NIST Gauge Block Measurement Assurance Program

The NIST gauge block measurement assurance program (MAP) has three parts. The first is a continual interferometric measurement of the master gauge blocks under process control supplied by control charts on each individual gauge block. As a measurement history is built up for each block, the variability can be checked against both the previous history of the block and the typical behavior of other similar sized blocks. A least squares fit to the historical data for each block provides estimates of the current length of the block as well as the uncertainty in the assigned length. When the master blocks are used in the mechanical comparison process for customer blocks, the interferometric uncertainty is combined in quadrature with the Intercomparison random error. This combined uncertainty is reported as the uncertainty on the calibration certificate.

The second part of the MAP is the process control of the mechanical comparisons with customer blocks. Two master blocks are used in each calibration and the measured difference in length between the master blocks is recorded for every calibration. Also, the standard deviation derived from the comparison data is recorded. These two sets of data provide process control parameters for the comparison process as well as estimates of short and long term process variability.

The last test of the system is a comparison of each customer block length to its history, if one exists. This comparison had been restricted to the last calibration of each customer set, but since early 1988 every customer calibration is recorded in a database. Thus, in the future we can compare the new calibration with all of the previous calibrations. Blocks with large changes are re-measured, and if the discrepancy repeats the customer is contacted to explore the problem in more detail. When repeated measurements do not match the history, usually the block has been damaged, or even replaced by a customer. The history of such a block is then purged from the database.

4.4.1 Establishing Interferometric Master Values

The general practice in the past was to use the last interferometric calibration of a gauge block as its "best" length. When master blocks are calibrated repeatedly over a period of years the practice of using a few data points is clearly not the optimum use of the information available. If the blocks are stable, a suitable average of the history data will produce a mean length with a much lower uncertainty than the last measurement value.

Reviewing the calibration history since the late fifties indicates a reasonably stable length for most gauge blocks. More information on stability is given in section 3.5. The general patterns for NIST master gauge blocks are a very small constant rate of change in length over the time covered. In order to establish a "predicted value," a line of the form:

$$L_t = L_o + K_1(t-t_o) \qquad (4.6)$$

is fit to the data. In this relation the correction to nominal length (L_t) at any time (t) is a function of the correction (L_o) at an arbitrary time (t_o), the rate of change (K_1), and the time interval ($t-t_o$).

When the calibration history is short (occasionally blocks are replaced because of wear) there may not be enough data to make a reliable estimate of the rate of change K_1. We use a few simple rules of thumb to handle such cases:

1. If the calibration history spans less than 3 years the rate of change is assumed to be zero. Few blocks under a few inches have rates of change over 5 nm/year, and thus no appreciable error will accumulate.

2. A linear least squares fit is made to the history (over 3 years) data. If the calculated slope of the line does not exceed three times the standard deviation of the slope, we do not consider the slope to be statistically significant. The data is then refitted to a line with zero slope.

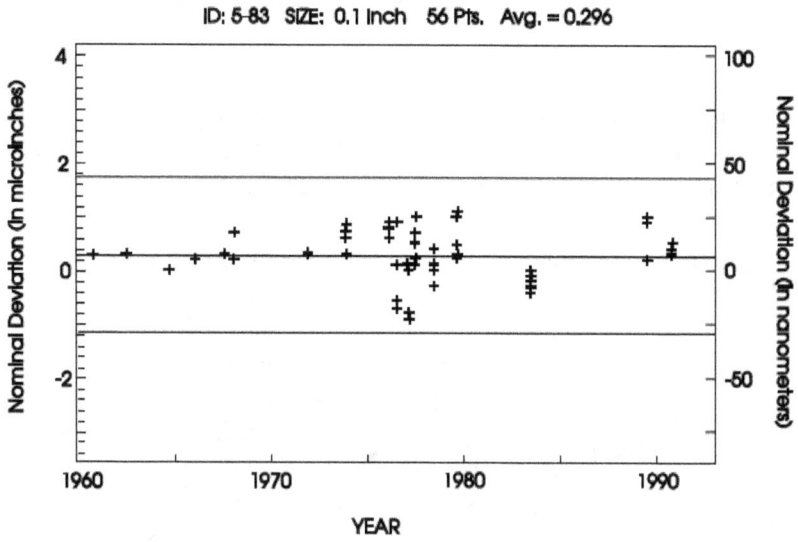

Figure 4.1a. Interferometric history of the 0.1 inch steel master block showing typical stability of steel gauge blocks.

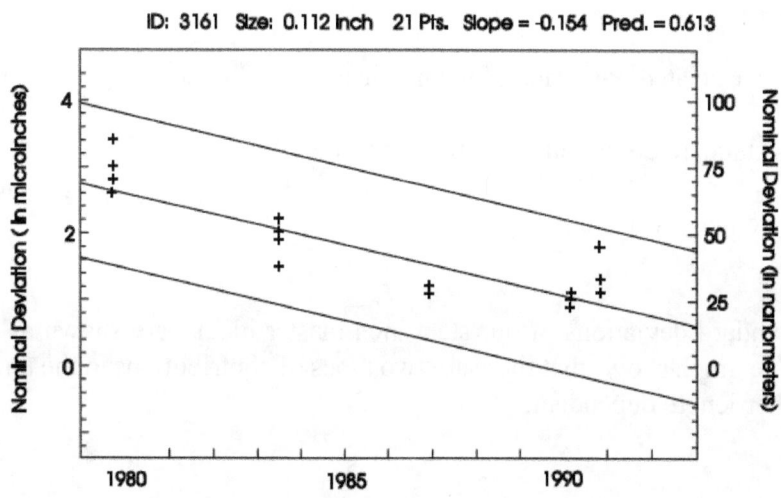

Figure 4.1b. Interferometric history of the 0.112 inch steel master block showing a predictable change in length over time.

Figure 4.1a shows the history of the 0.1 inch block, and it is typical of most NIST master blocks. The block has over 20 years of history and the material is very stable. Figure, 4.1b, shows a block which has enough history to be considered unstable; it appears to be shrinking. New blocks that show a non-negligible slope are re-measured on a faster cycle and if the slope continues for over 20 measurements and/or 5 years the block is replaced.

The uncertainty of predicted values, shown in figure 4.1 by lines above and below the fitted line, is a function of the number of points in each collection, the degree of extrapolation beyond the time span encompassed by these points, and the standard deviation of the fit.

The uncertainty of the predicted value is computed by the relation

$$3\sigma_t = 3\sigma \sqrt{\frac{1}{n} + \frac{(t-t_a)^2}{\sum(t_i-t_a)^2}} \qquad (4.7)$$

where n = the number of points in the collection;

t = time/date of the prediction;

t_a = average time/date (the centroid of time span covered by the measurement history, that is $t_a = \Sigma t_i/n$);

t_i = time/date associated with each of the n values;

σ = process standard deviation (s.d. about the fitted line);

σ_t = s.d. of predicted value at time t.

In figure 4.2 the standard deviations of our steel inch master blocks are shown as a function of nominal lengths. The graph shows that there are two types of contributions to the uncertainty, one constant and the other length dependent.

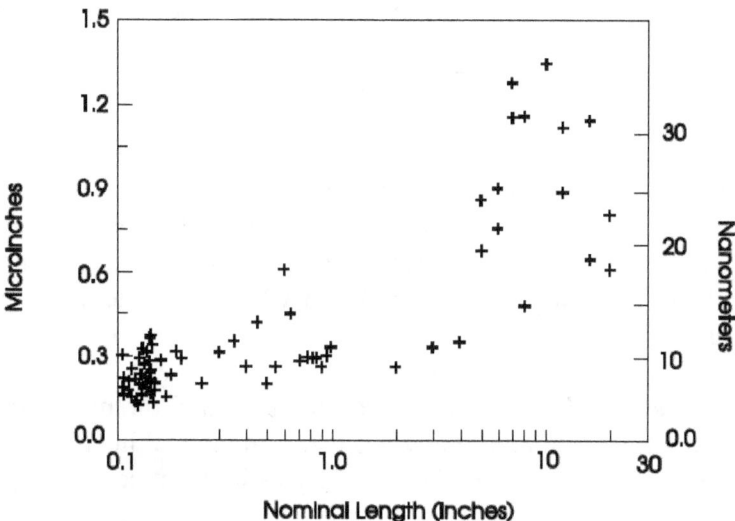

Figure 4.2. Plot of standard deviations of interferometric length history of NIST master steel gauge blocks versus nominal lengths.

The contributions that are not length dependent are primarily variations in the wringing film thickness and random errors in reading the fringe fractions. The major length dependent component are the refractive index and thermal expansion compensation errors.

The interferometric values assigned to each block from this analysis form the basis of the NIST master gauge block calibration procedure. For each size gauge block commonly in use up to 100 mm (4 in.), we have two master blocks, one steel and one chrome carbide. Blocks in the long series, 125 mm (5 in.) to 500 mm (20 in.) are all made of steel, so there are two steel masters for each size. Lengths of customer blocks are then determined by mechanical comparison with these masters in specially designed sequences.

4.4.2 The Comparison Process

There are currently three comparison schemes used for gauge block measurements. The main calibration scheme uses four gauge blocks, two NIST blocks and two customer blocks, measured six times each in an order designed to eliminate the effects of linear drift in the measurement system. For gauge blocks longer than 100mm (4 in.) there are two different schemes used. For most of these long blocks a procedure using two NIST blocks and two customer blocks measured four times each in a drift eliminating design is used. Rectangular blocks over 250 mm are very difficult to manipulate in the comparators, and a reduced scheme of two NIST masters and one customer block is used.

Comparators and associated measurement techniques are described in chapter 5. In this chapter, only the measurement assurance aspects of processes are discussed. The MAP for mechanical comparisons consists of a measurement scheme, extraction of relevant process control parameters, process controls, and continuing analysis of process performance to determine process uncertainty.

4.4.2.1 Measurement Schemes - Drift Eliminating Designs

Gauge block comparison schemes are designed with a number of characteristics in mind. Unless there are extenuating circumstances, each scheme should be drift eliminating, provide one or more control measurements, and include enough repeated comparisons to provide valid statistics. Each of these characteristics is discussed briefly in this section. The only scheme used at NIST that does not meet these specifications is for long rectangular blocks, where the difficulty in handling the blocks makes multiple measurements impractical.

In any length measurement both the gauge being measured and the measuring instrument may change size during the measurement. Usually this is due to thermal interactions with the room or operator, although with some laser based devices changes in the atmospheric pressure will cause apparent length changes. The instrumental drift can be obtained from any comparison scheme with more measurements than the number of unknown lengths plus one for drift, since the drift rate can be included as a parameter in the model fit. For example, the NIST comparisons include six measurements each of four blocks for a total of 24 measurements. Since there are only three unknown block lengths and the drift, least squares techniques can be used to obtain the best estimate of the lengths and drift. It has been customary at NIST to analyze the measurements in pairs, recording the difference between the first two measurements, the difference between the next two measurements, and so on. Since some information is not used, the comparison scheme must meet specific criteria to avoid increasing the uncertainty of the calibration. These designs are called drift eliminating designs.

Besides eliminating the effects of linear instrumental drift during the measurements, drift eliminating designs allow the linear drift itself to be measured. This measured drift can be used as a process control parameter. For small drift rates an assumption of linear drift will certainly be adequate. For high drift rates and/or long measurement times the assumption of linear drift may not be true. The length of time for a measurement could be used as a control parameter, but we have found it easier to restrict the measurements in a calibration to a small number which reduces the interval and makes recording the time unnecessary.

As an example of drift eliminating designs, suppose we wish to compare two gauge blocks, A and B. At least two measurements for each block are needed to verify repeatability. There are two possible measurement schemes: [A B A B] and [A B B A]. Suppose the measurements are made equally spaced in time and the instrument is drifting by some Δ during each interval. The measurements made are:

Measurement	Scheme ABAB	Scheme ABBA	
m_1	A	A	(4.8)
m_2	$B + \Delta$	$B + \Delta$	
m_3	$A + 2\Delta$	$B + 2\Delta$	
m_4	$B + 3\Delta$	$A + 3\Delta$	

If we group the measurements into difference measurements:

			Scheme ABAB	Scheme ABBA	
$y_1 =$	$m_1 - m_2$	$=$	$A - (B + \Delta)$	$A - (B + \Delta)$	(4.9a)
$y_2 =$	$m_3 - m_4$	$=$	$(A+2\Delta)-(B+3\Delta)$	$(B + 2\Delta)-(A+3\Delta)$	(4.9b)

Solving for B in terms of A:

$$B = \quad A - (y_1+y_2)/2 - \Delta \qquad A - (y_1-y_2)/2 \qquad (4.10)$$

In the first case there is no way to find B in terms of A without knowing Δ, and in the second case the drift terms cancel out. In reality, the drift terms will cancel only if the four measurements are done quickly in a uniformly timed sequence so that the drift is most likely to be linear.

A general method for making drift eliminating designs and using the least squares method to analyze the data is given in appendix A.

The current scheme for most gauge block comparisons uses two NIST blocks, one as a master block (S) and one as a control (C), in the drift eliminating design shown below. Two blocks, one from each of two customers, are used as the unknowns X and Y. The 12 measured differences, y_i, are given below:

$$
\begin{aligned}
y_1 &= S - C \\
y_2 &= Y - S \\
y_3 &= X - Y \\
y_4 &= C - S \\
y_5 &= C - X \\
y_6 &= Y - C \\
y_7 &= S - X \\
y_8 &= C - Y \\
y_9 &= S - Y \\
y_{10} &= X - C \\
y_{11} &= X - S \\
y_{12} &= Y - X
\end{aligned}
\qquad (4.11)
$$

$$\sigma_w^2 = \sum_{i=0}^{12} (\Delta Y_i)^2 / 8 \tag{4.12}$$

The calibration software allows the data to be analyzed using either the S block or the C block as the known length (restraint) for the comparison. This permits customer blocks of different materials to be calibrated at the same time. If one customer block is steel and the other carbide, the steel master can be used as the restraint for the steel unknown, and the carbide master can be used for the carbide unknown. The length difference (S-C) is used as a control. It is independent of which master block is the restraint.

There is a complete discussion of our calibration schemes and their analysis in appendix A. Here we will only examine the solution equations:

$S = L$ (reference length) (4.13a)

$C = (1/8) (-2y_1 + y_2 + 2y_4 + y_5 - y_6 - y_7 + y_8 - y_9 - y_{10} + y_{11}) + L$ (4.13b)

$X = (1/8) (-y_1 + y_2 + y_3 + y_4 - y_5 - 2y_7 - y_9 + y_{10} + 2y_{11} - y_{12}) + L$ (4.13c)

$Y = (1/8) (-y_1 + 2y_2 - y_3 + y_4 + y_6 - y_7 - y_8 - 2y_9 + y_{11} + y_{12}) + L$ (4.13d)

$\Delta = (-1/12)(y_1+y_2+y_3+y_4+y_5+y_6+y_7+y_8+y_9+y_{10}+y_{11}+y_{12})$ (4.13e)

The standard deviation is calculated from the original comparison data. The value of each comparison, $<y_i>$, is calculated from the best fit parameters and subtracted from the actual data, y_i. The square root of the sum of these differences squared divided by the square root of the degrees of freedom is taken as the (short term or within) standard deviation of the calibration:

For example,

$\Delta y_{12} = (C - B) - y_{12}$ (4.14a)

$= (1/8) (y_2 - 2y_3 + y_5 + y_6 + y_7 - y_8 - y_9 - y_{10} - y_{11}) - y_{12}$ (4.14b)

These deviations are used to calculate the variance.

The standard deviation (square root of the variance), a measure of the short term variation of the comparison process, is recorded for each calibration in the MAP file, along with other pertinent data from the calibration. These data are later used to determine the value of one control parameter, the short term variation, as described in the next section.

4.4.2.2 Control Parameter for Repeatability

There are a number of ways to monitor the comparator process. The simplest is to compare the variance of each calibration to an accepted level based on the history. This accepted level (average variance from calibration history) is characteristic of the short term repeatability of the comparator process, and is called the "within" variance.

In our system the within variance is used for two purposes. One is as a control point for the calibration, as discussed below. The second is that being a measure of short term variation of the entire calibration process, it is used to calculate the process uncertainty, a part of the total uncertainty reported to the customer.

In our system the variance for each calibration is recorded in the MAP file. Each year the last three years of data are analyzed to update the process control parameters used for the next year. Figure 4.3 shows a typical MAP file data plot.

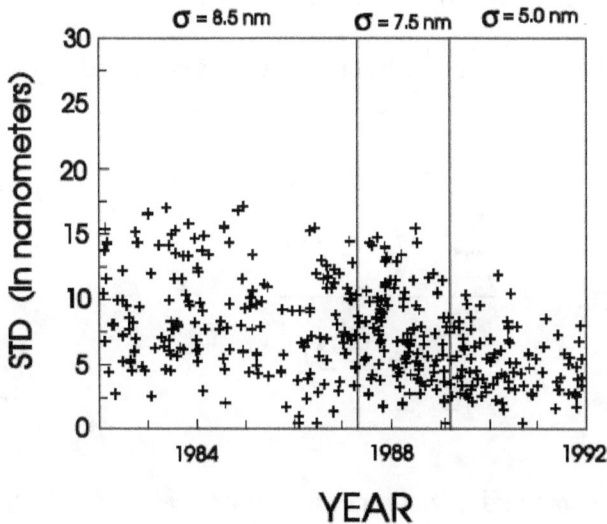

Figure 4.3. Variation of the short term standard deviation with time for the 0.1 inch size.

The most prominent characteristic of the graph is the continuing reduction of the standard deviation since 1981. The installation of new comparators during 1987 and the replacement of one of the old master sets with a new chrome carbide set in 1989 have resulted in significant reductions in the process uncertainty. When the current process parameter, σ_w, is calculated, only the data since the last process change (new master set in 1989) will be used.

If the calibration variance was independent of the block length only one value of the control parameter would be needed, and the average of the last few years of data from all of the calibrations could be used as the control point. The variance is, however, made up of many effects, some length dependent and some not.

Some of the non-length dependent part of the variance is due to electronic noise in the comparator. Another source is the contact point of the comparator sampling slightly different points on the surface of the block each time. Due to variations in flatness, parallelism and probe penetration into the surface, different lengths are measured.

The length dependent part of the variance is due to factors such as thermal variations in the blocks. The variation caused by this factor is not so strong that every block size must be considered separately. For example, the effects of thermal variations for 1 mm and 1.1 mm blocks should be nearly identical, although the same thermal variation on the 5 mm and 50 mm blocks will be measurably different. By separating the blocks into small groups of nearly the same size the number of measurements is increased, making the determination of the variance more accurate for the group than it would be if done on an individual basis.

Each group is made of about 20 sizes, except for blocks over 25 mm (1 in) where each size is analyzed separately. The groupings in current use are shown in table 4.4.

For each block size the average variance for the last three years is calculated. A weighted average of the variances from individual blocks is calculated, as shown in eq (4.4).

$$\sigma_p^2 = \frac{\sum_{j=1}^{m}(n_j - 2)S_j^2}{\sum_{j=1}^{m}(n_j - 2)} \tag{4.15}$$

In the above equation:

σ_p is the pooled standard deviation of the group,

m is the number of sizes in the group,

n_j is the number of measurements of size j,

s_j is the average standard deviation calculated from the MAP file for size j.

The within, σ_w is calculated for each group.

Table 4.4. Current Groupings for Calculating σ_w.

Group Number	Sizes			Units
1	0.010	to	0.09375	inch (thins)
2	0.100	to	0.107	inch
3	0.108	to	0.126	inch
4	0.127	to	0.146	inch
5	0.147	to	0.500	inch
6*	0.550	to	4.000	inch
7*	5.000	to	20.00	inch
8	0.0101	to	0.019	inch (.010 thins)
9	0.0201	to	0.029	inch (.020 thins)
10	0.0501	to	0.059	inch (0.050 thins)
11	0.200	to	0.2001	inch (step blocks)
12	unassigned			
13	0.10	to	0.95	millimeter (thins)
14	1.00	to	1.09	millimeter
15	1.10	to	1.29	millimeter
16	1.30	to	1.49	millimeter
17	1.50	to	2.09	millimeter
18	2.10	to	2.29	millimeter
19	2.30	to	2.49	millimeter
20	2.50	to	10	millimeter
21	10.5	to	20	millimeter
22	20.5	to	50	millimeter
23*	60	to	100	millimeter
24*	125	to	500	millimeter

* NOTE: for sizes over 1 inch and 50 mm each block history is analyzed individually, the groupings here are merely to simplify the data file structure.

4.4.2.3 Control test for the variance:

To test if the variance of an individual calibration is consistent with the short term variance derived from the calibration history we use the "F-test." The F-Test compares the ratio of these two variances to the F distribution, or equivalently the ratio of the standard deviations squared (variance) to the F distribution. Since the variance is always positive it is a one sided test:

$$\sigma^2_{cal}/\sigma^2_w < F_{1-\alpha}(k-1, v) \qquad (4.16)$$

where σ_w is the historical (pooled) short term group standard deviation, σ_{cal} is the standard deviation from the calibration, α is the upper percentage point of the F distribution with k-1 degrees of freedom in the numerator and ν degrees of freedom in the denominator. Using the 12 comparison calibration scheme, the number of degrees of freedom is 8, i.e. 12 measurements minus the number of unknowns (S,C,X,Y,Δ) plus one for the restraint. The number of degrees of freedom in finding σ_w from the history is generally a few hundred, large enough to use infinity as an approximation. The α used is 0.01. Using these parameters the control test becomes

$$\sigma^2_{cal}/\sigma^2_w < 2.5 \qquad (4.17)$$

If this condition is violated the calibration fails, and is repeated. If the calibration fails more than once the test blocks are re-inspected and the instrument checked and recalibrated. All calibrations, pass or fail, are entered into the history file. A flag is appended to the record to show that an F test failed.

4.4.2.4 Control parameter (S-C)

Control parameter σ_w will monitor certain aspects of the calibration, but not necessarily the accuracy. That is, measurements can have good repeatability and still be wrong. It is possible, for example, that the comparator scale calibration could be completely wrong and the data could still be repeatable enough that the variance (as measured by σ_w) would be in control. Our second control parameter samples the accuracy of the system by monitoring, for each size, the length difference between two NIST master blocks (S-C). No matter which master block is taken as the restraint this difference is the same.

Consider the measurement sequence shown in the previous sections. The customer block lengths X and Y, as well as the length of one of our master blocks are determined by a model fit to the comparison data using the known length of the other NIST master block (the restraint). The idea of this test is that while the customer block length is not known, the length of our control block is known and if the calibration process produces the correct length for the control block it probably has also produced the correct length for the customer block.

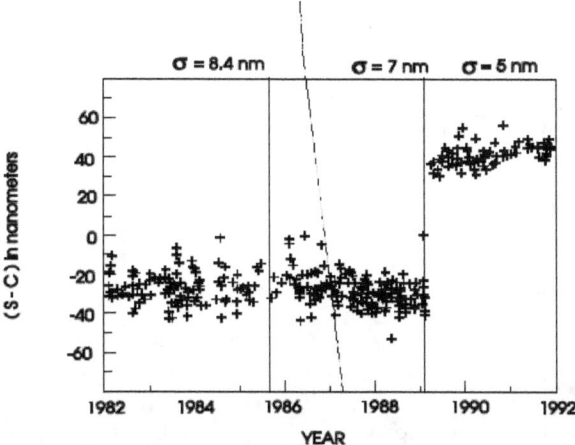

figure 4.4. (S-C) data from the MAP files for the 0.1in. blocks from 1982 to 1992. Note the process change in 1988-89 when the "C" block was replaced.

A plot, for the 0.1" size, of all of the S-C measurements from 1982 to the end of 1992 is shown in figure **4.4**. There are three distinct control regions shown in the graph by vertical lines. The first, which extends from 1982 to the middle of 1987 shows the performance of the system when we used two steel master blocks and our old (circa 1974) comparators. During the next period, until the beginning of 1990, we used new comparators with modern electronics. The higher repeatability of the new system resulted in a slightly smaller spread in the (S-C) values. In 1989 we replaced the "C" set of steel blocks with chrome carbide, resulting in a shift in the (S-C) values. When we analyze the MAP data to get the accepted value of S-C to use in the statistical test we obviously ignore all data before 1989.

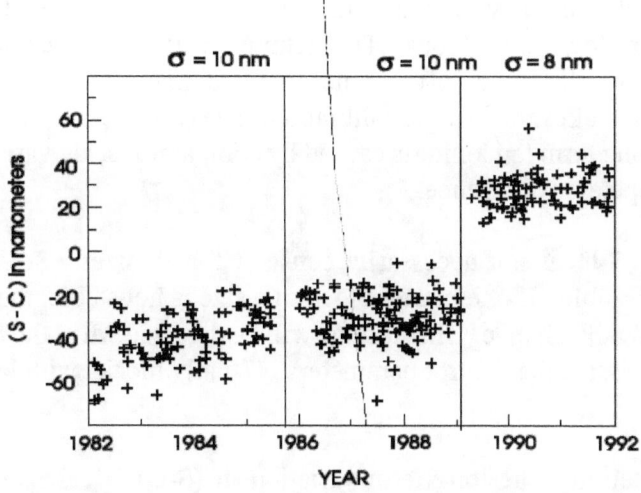

Figure 4.5. MAP data for 1.0 in. calibrations from 1982 to 1992. Note that the spread is somewhat larger than that of the 0.1 in size data in figure 4.4 because of temperature related uncertainties. The C block was replaced in 1989.

Figures 4.5 and 4.6 show the corresponding S-C graphs for the 1.0 in. and 10 in. sizes. The most prominent feature is that the data spread gets larger with length, a reflection of thermal variations during the calibrations. The 10 in. blocks show no process jump because the long block comparator and master blocks have not been changed during the time period shown.

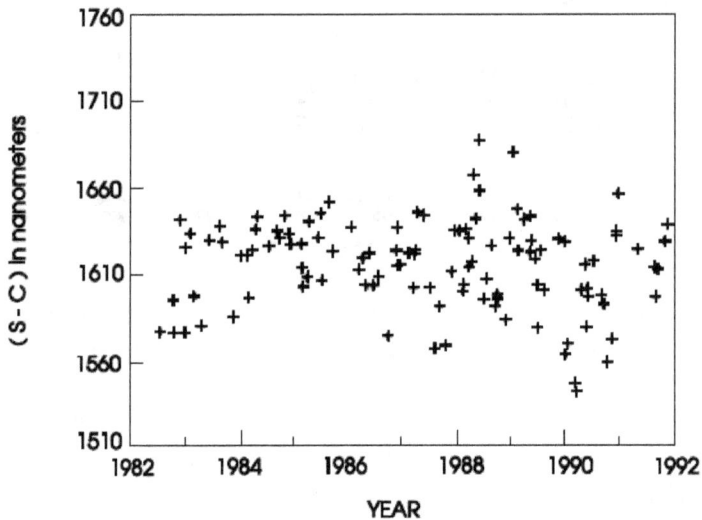

Figure 4.6. MAP data for 10 in. calibrations from 1982 to 1992. There is no process change for this size because the long block comparator and the two master blocks have not changed during the time period shown.

These charts provide a wealth of information about the calibration. Since the control block is treated exactly as the customer blocks, the variation in the control block length is identical to long term variation in the customer block calibrations. The within uncertainty, calculated from deviations from the model fit for each calibration only samples the random errors within characteristic times (less than 1 or 2 minutes) it takes for a single calibration. The (S-C) history samples both the short term variations and the long term randomness caused by comparator scale variations, thermometer calibration changes, or operator differences.

The long term value of (S-C) is not necessarily constant for all sizes. Some gauge blocks, as discussed earlier, are not stable. If one master block of a size is not stable, the (S-C) recorded for each calibration will gradually change. This change will be detected when the history for each size is analyzed each year to update the control parameters. Usually blocks with large instabilities are replaced.

Because of possible instability, the long term variation in (S-C) is calculated by the standard deviation of a linear fit to the (S-C) history data collected from past calibrations. Data for each size is fit to the form

$$L_i = m_i T + I_i \qquad (4.19)$$

where L_i is the length of the i'th block, m_i is the best fit slope, T is the time of the measurement, and I_i is the best fit intercept. The standard deviation of the slope, σ_{slope}, is determined by the fit procedure. If the block has a "significant" drift as determined by the test:

$$m_i > 3\sigma_{slope} \qquad (4.20)$$

then the value of the control parameter is taken as the fit value from equation 4.19 when T is equal to half way till the next update. If the slope is not significant the slope is assumed to be zero, and the data is simply averaged. The standard deviation of the data from the average is then determined.

These deviations are pooled for every block in a group and the pooled standard deviation, σ_t, is determined. This σ_t is recorded in the calibration master file to be used in the uncertainty calculation reported to the customer.

Finally, as discussed earlier short term variability is characterized by the within standard deviation, σ_w recorded at each calibration. Long term variability is characterized by σ_t calculated from the (S-C) history data. These two numbers can be compared and any non-negligible difference can be attributed to uncorrected long term variations or errors in the system. Based on the size of these uncorrected errors it may or may not be economically useful to search for their sources and reduce their size.

4.4.2.5 Control test for (S-C), the Check Standard

The length difference between the S and C blocks of each size are called check standards, and they monitor process accuracy. The control test for accuracy is based on a comparison of the (S-C) value of the current calibration with the accepted value of (S-C) derived from the (S-C) mechanical comparison data that has been accumulated for several years. This comparison is the second control test for the calibration, and the historically accepted mechanical value of (S-C) is the control parameter.

To be in statistical control, the current calibration must be consistent with the previous population of measurements. The t-test compares the current measurement of (S-C) to the accepted (S-C) value using the standard deviation derived from previous calibrations, σ_w, as a scale. If the difference between the current and accepted value is large compared to this scale the measurement system is judged to be out-of-control.

The actual test statistic is:

$$\frac{(Measured - Accepted)}{\sigma_w} < t_{\alpha/2}(\nu) \qquad (4.21)$$

Where ν is the number of degrees of freedom in the accepted total standard deviation, and α is the significance level. At NIST we use the 3σ, or 0.3 % level in calculating the total error, and therefore α=.003 is used in the t-test. The total number of measurements used to calculate the accepted total standard deviation is usually over 200, and we can use infinity as an approximation of v.

$$\frac{(Measured - Accepted)}{\sigma_w} < 2.62 \qquad (4.22)$$

With these assignments a t-test critical value of 2.62 is used. As a comparison, if v were chosen lower, due to having only a short calibration history, the changes would be small. For v=120 and 60 the critical values are only increased to 2.77 and 2.83 respectively.

If this condition is violated the calibration fails, and is repeated. If the calibration fails more than once the test blocks are re-inspected and the instrument checked and recalibrated. All calibrations, pass or fail, are entered into the history file. A flag is appended to the record to show that the t-test failed.

4.4.2.6 Control test for drift

As a secondary result of analyzing comparison data using a drift eliminating design, linear drift is measured and can be used as another control parameter. If the drift shows nonlinear behavior in time, the design will not completely eliminate the drift, and the answers will be suspect. However, nonlinear drift will accentuate the deviations from the best line fit of the data and the calibration will fail the F-test. Thus a control based on the drift rate would be redundant. In either case, because of the environmental control used in our lab (0.1 °C) drift has never been significant enough to consider such a test.

4.4.3 Calculating the total uncertainty

In the general method for combining uncertainties discussed in section 4.3.6 total uncertainty is the root-sum-of-squares of the uncertainty components times a coverage factor k=2.

For NIST calibrations the total error is made up of random error from the master block interferometric history and long term random error of the comparator process derived from control data. Other sources of uncertainty are the phase correction assigned to each block in the interferometric measurements and the temperature difference between the customer block and the NIST master block. For blocks over 50 mm the temperature gradients and non-negligible sources of uncertainty..

The long term random error of the comparison process is calculated from the (S-C) data collected in the MAP file. Since each block, master and unknown, is handled and measured in exactly the same manner, the variance of the difference (S-C) is a reasonable estimator of the long term variance of the customer calibration (S-X) or (C-X). The major difference is that since the S and C blocks are of different materials, the variation will be slightly larger than for customer calibrations, for which

(Standard-X) is always derived from the master block of the same material as the customer block. Although we expect this difference is small, we have not collected enough customer data to analyze the difference, and our total uncertainty must be considered slightly inflated.

The total (long term) standard deviation (σ_{tot}) is calculated by groups, in much the same manner as σ_w. One additional step is included because of the possible instability in the length of one of the master blocks. The (S-C) history of each size is fit to a straight line and the deviations from the line are pooled with the deviations from the other sizes in the group, as shown in eq (4.15).

$$\sigma_{tot} = \sqrt{\frac{\sum_{j=1}^{m} D_j^2}{(m-2)}} \quad (4.23)$$

In the above equation:

σ_{tot} is the pooled long term standard deviation of the group,

m is the number of measurements of all sizes in the group,

D_j is the deviation from the best fit to the (S-C) history data for each size.

As discussed by Cameron [5], the random error for calibration schemes using incomplete block designs is a combination of the short and long term repeatability. In these cases short term repeatability is the average of the standard deviations of each calibration recorded in the calibration history, σ_w, and the long term repeatability is σ_{tot}. For complete block designs only σ_{tot} is needed.

The interferometric uncertainty is essentially a random error. If each block were measured only a few times on a single date the pressure, temperature and humidity calibrations used to make corrections for the index of refraction and thermal expansion would be sources of systematic error. However, each time our master blocks are re-measured (3 year intervals) the pressure, temperature and humidity sensors are recalibrated. After the blocks have been measured using three or more sensor calibrations, errors of the environmental sensors have been sampled and the variations due to them have been absorbed into the gauge block calibration data.

The interferometric phase correction has a fairly high uncertainty because of the difficulty of the measurements. In addition, phase measurement is much too difficult to make on each block and so only a few blocks from each manufacturer are tested and the average of these measurements are assigned to all blocks from the same manufacturer. Estimated uncertainty in the phase correction is around 8 nm (0.3 μin).

We have measured temperature gradients on all of our comparators. On short block comparators

(<100 mm) we have found gradients up to 0.030 °C between blocks. Since we do not have extensive measurements we consider this as a rectangular distribution of total width 0.030 °C. Using the rules of NIST Technical Note 1297 we divide the width by the square root of 3, giving an uncertainty of 0.020 °C. The long block comparators are partially enclosed which reduces the gradients below 0.010 °C. The standard uncertainty from these gradients is only 0.005 °C.

4.5 Summary of the NIST Measurement Assurance Program

Two major goals of the gauge block MAP program are to provide a well-controlled calibration process to ensure consistently accurate results, and to quantitatively characterize the process so that a rational estimate of process accuracy can be made. Process control is maintained with three tests and data collected from the calibrations is used for both the tests and the estimation of process uncertainty.

In the first test interferometric history of each master block is continually updated and checked for consistency. Suspicious values are re-measured. As the number of measurements in each history increase, the dimensional stability of each block can be determined. If the block is changing length, the projected length at the time of the comparison measurement can be used, or the block can be replaced. The standard deviation of the history data for each block is used as an estimate of the long term random uncertainty of the master block value.

In the second test the standard deviation of each mechanical comparison is checked against the average value of the standard deviations from previous calibrations. This average standard deviation for an individual calibration is called the "within" or short term standard deviation (σ_w). Since the error sources for similar size blocks are similar, we pool the historical data for groups of similar sized blocks to increase statistical precision. The standard deviation from each calibration is tested against the accepted σ_w by the formula:

$$\sigma^2_{cal}/\sigma^2_w < 2.5 \qquad (4.24)$$

In the final test each mechanical comparison includes a check standard as well as the master block. The difference between these two NIST blocks, (S-C), is tested against an accepted value obtained by a linear fit to the recorded (S-C) values for the last few years. The difference between the value obtained in the calibration and the accepted value is measured against the expected standard deviation of the measurement, σ_w, using the formula:

$$\frac{(Measured - Accepted)}{\sigma_w} < t_{\alpha/2}(\nu) \qquad (4.25)$$

The uncertainty reported to the customer is a combination of the type A and type B uncertainties. The major components of uncertainty are the standard deviation of the interferometric history of the master block used in the calibration, the long term standard deviation of the mechanical comparison process, σ_{tot}, and the estimated type B uncertainty. The primary components of type B uncertainty are the phase correction applied during interferometric calibrations of our master gauge blocks and the temperature gradients between blocks during the comparison. All of these errors are combined in quadrature and multiplied by a coverage factor (k) of two, thus arriving at the reported uncertainty.

$$Total\ Uncertainty\ =\ k\sqrt{\Sigma A^2 + \Sigma B^2} \tag{4.26}$$

5. The NIST Mechanical Comparison Procedure

5.1 Introduction

Techniques described here for intercomparing gauge blocks with electro-mechanical gauge block comparators are those used at NIST and are the results of considerable experimental work, statistical design and process analysis. Some laboratories will find these techniques ideally suited to their needs while others may wish to modify them, but in either event the process control methods described here will promote the analysis of measurement requirements and the application of statistical principles.

Length values are assigned to unknown gauge blocks by a transfer process from blocks of known length. Briefly, the transfer process is a systematized intercomparison of 3 or 4 blocks of the same nominal size using an electro-mechanical comparator. One or two blocks are unknowns and two are standards. Such a process is ideally suited to comparing objects with nearly identical characteristics, such as gauge blocks, and the process provides the redundancy needed for statistical analysis.

We will also describe the corrections needed for the case where the blocks are not made of identical materials. The correction methods apply to all sizes but special precautions will be described for long gauge blocks where temperature effects predominate. When the master and unknown blocks are of the same material, which for our calibrations include both steel and chrome carbide blocks, no corrections are needed.

5.2 Preparation and Inspection

It is essential that the gauge blocks being calibrated, as well as the standards being used, be clean and free from burrs, nicks and corrosion if the calibration is to be valid. Therefore, it is necessary to institute an inspection, cleaning and deburring procedure as part of every calibration. Such a procedure will also protect the comparator working surface and the diamond styli.

5.2.1 Cleaning Procedures

Gauge blocks should be cleaned in an organic solvent such as mineral spirits. Solvents such as trichloroethylene or benzene should be avoided because of health risks involved with their use. Freons should not be used because of environmental hazards they pose. The bottom of the bath should be lined with a waffled mat of neoprene rubber to minimize scratches and burrs to the gauge block surfaces caused by grit or metal to metal contact.

Rectangular blocks should be immersed in the bath and a soft bristle brush used to remove the layer of protecting oil or grease from all surfaces. Lint-free towels should be used to dry the blocks. A second cleaning using ethyl alcohol is recommended to remove any remaining residue. Again, wipe with lint-free towels.

Square type blocks are cleaned in the same manner as the rectangular blocks except that the center hole needs special attention. The hole must be thoroughly cleaned and dried to prevent "bleeding out" of an oil film onto the contact surfaces and supporting platen when the block is positioned vertically. A .22 caliber gun cleaning rod makes an ideal instrument for cleaning the bore of blocks in the size range of 2 mm to 500 mm. A solvent moistened piece of lint-free paper towel can be pushed through the hole repeatedly until all visible residue is removed.

5.2.2 Cleaning Interval

It is recommended that the solvent and alcohol cleaning process be completed several hours prior to use because the cleaning process will disturb the gauge block thermal state. If blocks are to be joined together or to platens by wringing, recleaning is necessary prior to introducing a wringing film to ensure that no foreign substances are on the wringing surfaces that could corrode or mar the contact surfaces.

5.2.3 Storage

Cleaned blocks should be stored in trays covered with lint-free paper or on temperature equalization plates that are clean. A cover of lint-free towels should be used to secure the blocks from dust and other airborne solids and liquids. When blocks are not in daily use, they should be coated with any one of the many block preservatives recommended by the gauge block manufacturers. It is recommended that steel gauge blocks be coated to prevent rust when the work area relative humidity exceeds 50%.

Foodstuffs should be banned from all calibration areas, as acids and salts will corrode steel gauge surfaces. Hands should be washed to reduce corrosion on blocks when gloves or tongs cannot be used conveniently.

5.2.4 Deburring Gauge Blocks

Gauge blocks can often be restored to serviceable condition by stoning if the damage is not severe. Three different optically flat gauge block stones are available: black granite, natural Arkansas, or sintered aluminum oxide. All stones are used in the same manner, but are used for different purposes or on different materials.

The black granite deburring stone is used on steel blocks to push a burr of metal back into the area from which it was gouged. The natural Arkansas deburring stone is a white fine-grained stone that is used on steel blocks to remove burrs and nicks caused by everyday mishaps. It is especially effective on damaged edges of gauging faces. The sintered aluminum oxide stone is used to treat either steel or carbide gauge blocks. It is most effective on hard materials such as tungsten and chromium carbide.

The stone and block should both be cleaned with alcohol before stoning. Sweep the surfaces with a camel's hair brush; then slide the damaged surface of the block onto the stone, taking precautions that no foreign matter is trapped between the two surfaces, and with firm downward pressure the block is moved over the stone in either a circular or a back and forth motion until the metal is restored to its original placement. Generally, there is a lessening of drag between the two surfaces as the burr is reduced.

After stoning and cleaning a block, one can check the surface by wringing the restored surface to a fused silica flat. Looking through the back side of the flat, the interface between the block and flat should appear uniformly grey when it reflects white light. Any light areas or color in the interface indicates poor wringing caused by a burr that has not been fully reduced to the normal gauge surface.

Periodically, depending on use, the stone should be cleaned with a stiff natural fiber brush, soap and water to remove embedded pieces of metal and other foreign matter that contaminates the surface with use. Allow moisture absorbed by the stone surface to evaporate before using the stone.

5.3 The Comparative Principle

A comparator is a device for measuring differences. The unknown length of a block is determined by measuring the difference between it and a reference block of the same nominal size and then calculating the unknown length. Let L_c equal the comparator length when its indicator reads zero (the zero may be at the left end or the center of the scale). Then, by inserting the two blocks of length L_x (unknown) and L_r (reference) and reading numerical values x and r from the comparator scale, we get

$$L_x = L_c + x \qquad (5.1)$$
$$L_r = L_c + r \qquad (5.2)$$

Solving these equations for L_x:

$$L_x = L_r + (x-r). \qquad (5.3)$$

Note that the quantity (x-r) is the difference between the two blocks. If the blocks are made of different materials we add a quantity, δ, to each comparator reading to compensate for elastic deformation of the contact surfaces. We must also add a correction for the thermal expansion if the calibration temperature T is not 20 °C. The equation becomes

$$L_x = L_r + (x - r) + (\delta_x - \delta_r) + L(\alpha_r - \alpha_x)(t-20). \qquad (5.4)$$

The quantity $(\delta_x - \delta_r)$ in this equation is the deformation correction, C_p. α_x and α_r are the thermal expansion coefficients, and L the blocks' nominal lengths. Using C_t for this thermal correction the equation becomes

$$L_x = L_r + (x-r) + C_p + C_t. \qquad (5.5)$$

If two probes are used, one lower and one upper, there are two penetration corrections and C_p must contain a lower and upper component.

A key result of the comparative principle is that if the two blocks being compared are the same material, C_p and C_t are zero.

5.3.1 Examples

1. A single probe comparator is used to measure a 50 mm tungsten carbide gauge block with a 50 mm steel reference standard gauge block. The measurement data are:

Stylus force (top):	0.75 N
Stylus force (bottom):	0.25 N
Stylus tip material:	diamond
Stylus tip radius:	6 mm
L_r of reference at 20 °C	50.00060 mm
Temperature of blocks:	20.4 °C
Expansion coefficient:	
Steel	$11.5 \times 10^{-6}/$ °C
Tungsten carbide	$6 \times 10^{-6}/$ °C
Comparator reading on tungsten block	1.25 µm
Comparator reading on steel block	1.06 µm

The unknown length (L_x) at 20 °C can be found using the last equation and the penetration corrections from table 3.4:

	Penetration, µm (µin)
Steel	0.14 (5.7)
Tungsten Carbide	0.08 (3.2)

Then

$$L_x = 50.00060 + (1.25-1.06) \times 10^{-3} + 50 \times (11.5-6) \times 10^{-6} \times (20.4 - 20) + (0.08 - 0.14) \times 10^{-3} \quad (5.6a)$$

$$L_x = 50.00060 + \quad 0.00019 \quad + \quad 0.00011 \quad - \quad 0.00006 \quad (5.6b)$$

$$L_x = 50.00084 \text{ mm} \quad (5.6c)$$

2. A dual probe comparator is used to measure a 10 mm steel block with a 10 mm chrome carbide reference block. From the following data find the length of the steel block at 20 °C.

Stylus force:	0.75 N upper, 0.25 N lower
Stylus tip material:	diamond
Stylus tip radius:	6 mm
L_r of reference at 20 °C	9.99996 mm
Temperature of blocks:	23.6 °C
Expansion coefficient:	
Steel	$11.5 \times 10^{-6}/°C$
Chrome carbide	$8.6 \times 10^{-6}/°C$
Comparator reading on chrome block	0.00093 mm
Comparator reading on steel block	0.00110 mm

	penetration (0.25 N)	penetration (0.75 N)	TOTAL
Steel	.07 μm	0.14 μm	0.21 μm
Chrome Carbide	.06 μm	0.12 μm	0.18 μm

Solution:

measurements   thermal correction   penetration correction

$$L_x = 9.99996 + (.00110 - .00093) + 10 \times (8.6 - 11.5) \times 10^{-6} \times (23.6 - 20) + (.21 - .18) \times 10^{-3} \quad (5.7a)$$

$$L_x = 9.99996 + .00017 - .00010 + 0.00003 \quad (5.7b)$$

$$L_x = 10.00006 \text{ mm} \quad (5.7c)$$

In summary, deformation corrections can be easily and accurately applied if the operative parameters of probe tip radius and probe force are known and if the probe tip is spherical and remains undamaged. Extremely low forces should be avoided, however, because poor contact is a risk. It is also important to know that thermal conditions, if ignored, can cause far greater errors than deformation, especially for long lengths.

Currently the NIST comparator program uses steel master blocks for steel unknowns and chrome carbide masters for chrome or tungsten carbide unknowns. This allows the penetration correction to be ignored for all of our calibrations except for a very few sets of tungsten carbide and other materials that we receive. Because of the small number of gauge blocks sent for calibration it is not economically feasible to have master sets for every possible gauge block material. For these blocks the correction is minimized by the use of the master blocks with the most closely matched elastic properties.

5.4 Gauge Block Comparators

There are a number of suitable comparator designs and a typical one is shown schematically in figure 5.1. An upper stylus and a lower stylus contact the gauging faces of a block supported on an anvil. Each stylus is attached to a linear variable differential transformer (LVDT) core. The amplifier displays the difference between the two LVDT signals.

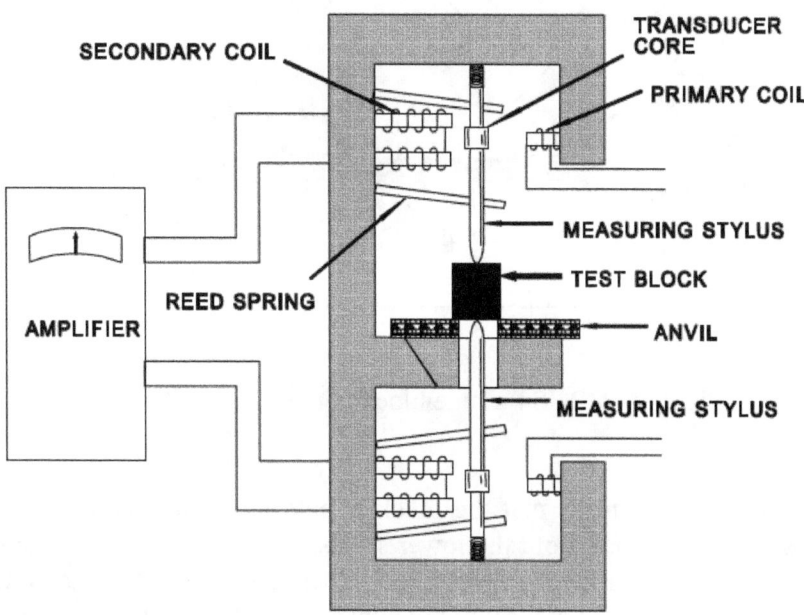

Figure 5.1. Schematic drawing of a gauge block comparator showing the component parts.

The NIST comparators have a more complex mechanical arrangement, shown in figure 5.2. The use of pivots allows measurement of the relative displacement of two measuring styli using only one LVDT. Since one of the main causes of variations in measurements is the LVDT repeatability, reducing the number of LVDTs in the system enhances repeatability.

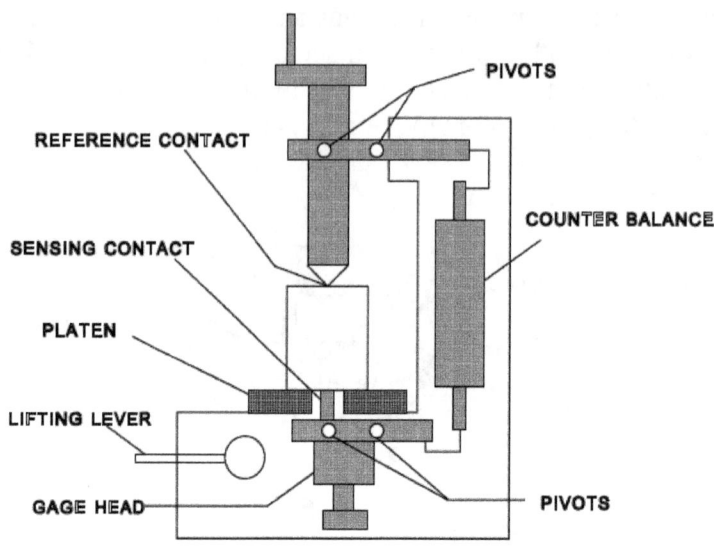

Figure 5.2. NIST gauge block comparator.

An important comparator feature is that the block is held up by the anvil (also called platen), removing the weight of the block from the lower stylus. The force of the gauging stylus on the bottom of the block is therefore the same for every block measured. Were this not the case the corrections needed for deformation at the point of contact would depend on the weight of the block as well as it's material. There are, however, systems that support the block on the lower stylus. In these cases a penetration correction which depends on the material, size and geometry of the block must be made for each block.

Since the LVDT range is small, the height of one of the sensors must be mechanically changed for each size gauge block. The highest feature on figure 5.2 is the crank used to rotate a fine pitch screw attached to the top stylus. For each size the top stylus is set and locked so that the LVDT output is on the proper scale. The lifting lever raises the top stylus from the block and locks the pivot system so that the stylus cannot crash into the gauging surface by mistake during the gross adjustment to the next size.

For the most precise measurements, it is important that the defined gauge point of a block be contacted by the comparator stylus. If the two gauging surfaces are not exactly parallel the measured block length will vary with the gauging point position. This is particularly important for long blocks, which tend to have less parallel gauging surfaces than short blocks. To ensure that contact will be made at the proper point a metal or plastic bar about 1/4 inch thick is fastened to the anvil behind the stylus and is positioned to stop the gauge blocks so the stylus will contact the gauge point. The stop used on our long block comparator is shown in figure **5.3**.

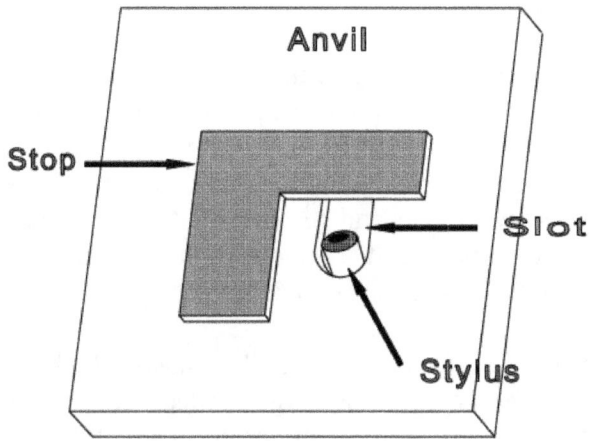

Figure 5.3. The mechanical stop is used to force the gauge block to a nearly identical position above the stylus for each measurement.

The LVDT signal (or difference between two LVDTs) is displayed on a meter graduated in length units. The analog output is also read by a digital voltmeter which, when triggered by a foot switch, sends the measured voltage to a microcomputer. The difference in length between two gauge blocks is obtained by inserting the blocks, one at a time, between the stylus tips and taking the difference between the two readings.

5.4.1 Scale and Contact Force Control

The comparator transducer must be calibrated to relate output voltage to probe displacement from some arbitrary zero. Our comparators are calibrated regularly using a set of 11 gauge blocks that vary in length from 0.200000 inches to 0.200100 inches in 10 microinch increments. Each calibration is recorded in a probe calibration file and the last 20 calibrations are available to the operator as a control chart with appropriate control limits. If a probe calibration is inconsistent with the previous calibrations the instrument is examined to find the cause of the discrepancy.

The NIST comparators for blocks less than 100 mm long use contact forces of 0.75 N (3 ounces) for the upper contact and 0.25 N (1 ounce) for the lower contact. The corresponding forces for the long gauge block comparator (over 100 mm) are 1.2 N (5 oz) and 0.75 N (3 oz). The force on both upper and lower contacts are checked daily using a calibrated, hand held, force gauge. If the force is found to be in error by more than 0.1 N, adjustments are made.

5.4.2 Stylus Force and Penetration Corrections

The penetration of the spherical stylus into the planar gauge block has already been presented in table 3.4. Values are correct to a few nanometers. Calculation of the value assumes a perfectly spherical surface with a nominal radius of curvature of 3 mm. This assumption is not always true.

Effects of deviations from the assumed probe geometry can be significant and have plagued comparison procedures for many years.

Usually the probe is not exactly spherical, even when new, because of the characteristics of diamond. The probe is manufactured by lapping, which if diamond were an isotropic media would produce a spherical form. Diamond, however, is harder in some directions than in others. The lapping process removes more material in the soft directions producing an ellipsoidal surface. Even worse, the diamond surface slowly wears and becomes flattened at the contact point. Thus, even though the calculation for deformation of a spherical-planar contact is nearly exact, the real correction differs because the actual contact is not a perfect sphere against a plane.

In the past these effects were a problem because all NIST master blocks were steel. As long as the customer blocks were also steel, no correction was needed. If, however, other materials were measured a penetration correction was needed and the accuracy of this correction was a source of systematic error (bias).

A gauge block set bias is determined from the history of each block. First, the yearly deviation of each block is calculated from its historical average length. Then the deviations of all the blocks are combined by years to arrive at the average change in the length of the set by year. This average should be close to zero, in fact if there are 81 blocks in the set the average will, if in control, fall within the limits $3\sigma/81$ or $\sigma/3$. The method of examining set averages is useful for many types of systematic errors and is more fully described in Tan and Miller [36].

Figure 5.4 shows the set bias for three customer gauge block sets having lengthy histories. Two sets, represented by squares are steel and the third, represented by asterisks, is chrome carbide.

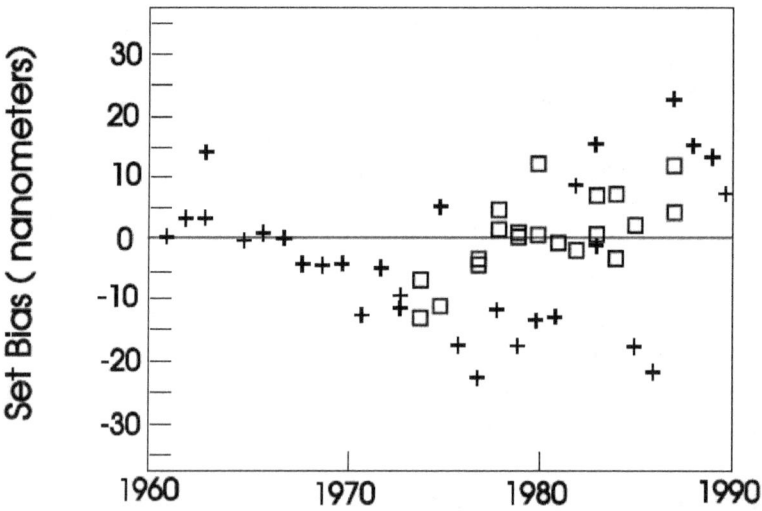

Figure 5.4. Bias of three gauge block sets since 1960. The squares are steel blocks and the crosses are chrome carbide blocks.

The two steel set calibration histories are considerably better than those for the chrome carbide. Also, before the early 1970's all calibrations were done by interferometry. The chrome carbide calibrations have degenerated since 1970 when mechanical comparisons were introduced as the calibration method.

In the past it was believed that carbide calibrations were not as accurate as steel and one extra microinch of uncertainty was added to carbide calibrations. The graph shows that this belief gained from experience was accurate. The NIST calibration procedures were changed in 1989 to eliminate this systematic error by having one set of master blocks made from chrome carbide. When steel blocks are measured the steel master is used as the reference block and when carbide blocks are measured the chrome carbide block is used as the reference. In each case the difference in length between the steel and carbide masters is used as the check standard.

5.4.3 Environmental Factors

A temperature controlled laboratory is essential for high precision intercomparisons. The degree of temperature control needed depends on the length of the blocks being compared, differences in thermal expansion coefficients among the blocks, and the limiting uncertainty required for the unknown blocks. At NIST, the gauge block calibration areas are specially designed laboratories with temperature control at 20.1 °C. Special precautions, described in this chapter, are taken for blocks over 100 mm.

5.4.3.1 Temperature Effects

The correction equation for thermal expansion is

$$\delta L/L = \alpha(20-T) \qquad (5.8)$$

Where L is the lenght of the block, δL the change in length due to thermal expansion, α is the thermal expansion coefficient, and (20-T) the difference between the measurement temperature and 20 °C.

When two blocks are compared, the length difference (m=S-X) between the blocks may need correction due to the difference in thermal expansion coefficients and temperatures of the blocks. If corrected the calibration result becomes

$$X_{cal} = S_{ref} - m + \delta L(\text{thermal}) \qquad (5.9)$$

$$\delta L/L = (\alpha_X - \alpha_S)(20 - T_S) + \alpha_X(T_S - T_X) \qquad (5.10)$$

Uncertainty of the calculated δL arises from uncertainty in α, ΔT, and the generally unmeasured temperature difference between S and X. An estimate of the upper limit of this uncertainty is:

$$\text{Uncertainty} = L(\alpha_X - \alpha_S)(\delta T_X) + L(\delta\alpha_S + \delta\alpha_X)(20 - T_S) + L\alpha_X(T_S - T_X) \quad (5.11)$$

where $\delta\alpha$ is the uncertainty in the thermal expansion coefficient and δT is the uncertainty in the temperature measurement.

We illustrate the effects with a simple example. Three 250 mm gauge blocks are being intercompared, a steel block is the reference master designated S, and one steel block and one chrome carbide block are designated X and Y. The thermal expansion coefficients and related uncertainties are:

$$S: (11.5 \pm 0.1) \times 10^{-6}/°C$$
$$X: (11.5 \pm 0.5) \times 10^{-6}/°C$$
$$Y: (8.4 \pm 0.5) \times 10^{-6}/°C$$

The blocks are measured at 20.5 °C with an uncertainty of 0.2 °C, and from previous practice it is known that there is the possibility of a temperature difference between blocks of up to 0.1 °C.

Length errors in three categories can be illustrated with these data: those caused by (1) temperature measurement uncertainty, $\alpha \delta T$; (2) gauge block expansion coefficient uncertainty, $L\delta\alpha(t-20)$; and (3) temperature differences between the reference and unknown blocks.

(1) The temperature measurement uncertainty of 0.2 °C together with the expansion coefficients of the unknown blocks yields:

$$\text{Uncertainty} = L(\alpha_X - \alpha_S)\delta T \quad (5.12)$$

(2) The uncertainty in the thermal expansion coefficients together with a measurement temperature different than 20 °C yields:

$$\text{Uncertainty} = L(\delta\alpha_S + \delta\alpha_X)|20 - T_S| \quad (5.15)$$

$$U(X \text{ or } Y) = 250 \times (0.1 \times 10^{-6} + 0.5 \times 10^{-6}) \times 0.5 = 0.00007 \text{ mm} \quad (3 \text{ μin}) \quad (5.16)$$

(3) If the master and unknown blocks are not at the same temperature, there will be an error depending on the size of the temperature gradient:

$$\text{Uncertainty} = L\,\alpha_X(T_S - T_X) \quad (5.17)$$

$$U(X) = 250 \times 11.5 \times 10^{-6} \times 0.1 = 0.0003 \text{ mm} \quad (12 \text{ µin}) \quad (5.18)$$

$$U(Y) = 250 \times 8.4 \times 10^{-6} \times 0.1 = 0.0002 \text{ mm} \quad (8 \text{ µin}) \quad (5.19)$$

The example shows that the uncertainty in the thermal expansion coefficient is not important if the temperature is kept close to 20 °C. The uncertainty in the temperature measurement is very important for blocks longer than 25 mm, and is critical for blocks longer than 100 mm. Finally, temperature differences between blocks are a serious matter and measures to prevent these gradients must be taken. Ways of controlling temperature effects are discussed in the next section.

5.4.3.2 Control of Temperature Effects

In the NIST comparison process it is assumed that the gauge blocks are all at the same temperature, i.e. the temperature of the stabilization plate or comparator anvil. A large error in the comparison process can be introduced by temperature gradients between blocks, or even within a single block. For example, a temperature difference of 0.5 °C between two 100 mm steel blocks will cause an error of nearly 0.6 µm (24 µin) in the comparison. Two causes of temperature differences between blocks are noteworthy:

1. Room temperature gradients from nearby heat sources such as electronic equipment can cause significant temperature differences between blocks even when they are stored relatively close to each other before comparison.

2. Blocks with different surface finishes on their non-gauging faces can absorb radiant heat at different rates and reach different equilibrium temperatures.

The magnitude of these effects is proportional to gauge block length.

A number of remedies have been developed. For short blocks (up to 50 mm) the remedies are quite simple. For example, storing the blocks, both standards and unknowns, on a thermal equalization plate of smooth surface and good heat conductivity close to the comparator but away from heat sources. Also, the use of insulated tweezers or tongs to handle the blocks and a systematic, rhythmic block handling technique in the comparison procedure to ensure identical thermal environment for each block. The use of a special measurement sequence called a drift eliminating design is discussed in section 5.6.

Precautions for long blocks (over 50 mm) are more elaborate. At NIST the comparator is enclosed in an insulated housing partially open on the front to allow handling the blocks with tongs. Blocks are stored on the comparator anvil and each block is wrapped in three layers of aluminized mylar to assure that each block has the same reflectivity to incident radiation. The insulation, although very thin, also limits the largest heat transfer mechanism, convection, during the measurement. This reduces thermal changes due to operator presence during calibration.

Laboratory lights remain off except for one lamp well removed from the comparator, but giving enough illumination to work by. The visible light is a secondary radiation problem, the infrared radiation from the lights is the primary problem. Heat sources are kept as far away from the comparator as possible and the comparator is located away from other activities where laboratory personnel might congregate. As a further precaution during intercomparisons the operator wears a cape of Mylar reflective film and heavy cotton gloves while handling the blocks with tongs. Finally, as with short blocks, the handling procedure is systematic, rhythmic and quick but not rushed. Rectangular blocks longer than 100 mm are measured using a reduced number of intercomparisons because of difficulty in handling this type of long block with its tendency to tip on its narrow base.

Temperature problems can be detected in long block intercomparisons by reversing the storage positions of the 4 blocks and repeating the intercomparison after suitable equalization time. Temperature gradients in the storage area will be revealed by a significant change in the relative lengths of the blocks. Still another method is to measure the temperature of each block with a thermocouple. A simple two junction thermocouple used with a nanovoltmeter can measure temperature differences of 0.001 °C or less. Tests should be made for gradients between blocks and internal gradients (top to bottom) within each block. When a gradient free environment is established a thermometer mounted on the comparator will be adequate for detecting unusual or gross changes.

Relative humidity is held below 50% to prevent corrosion of blocks and instruments. A full discussion of environmental requirements for calibration laboratories can be found in the ANSI standard B89.6.2.

5.5 Intercomparison Procedures

Once the blocks have been cleaned, as described in section 5.2 they are laid out on an aluminum soaking plate to allow all of the blocks to come to the same temperature. The blocks are laid out in 4 parallel rows in each size sequence, each row being a different set. The two NIST master blocks, S and C, are the top and second row respectively because each comparison scheme begins S-C. The customer sets, X and Y are set out as the third and fourth rows. It is not necessary to use this specific setup, but as in all facets of calibration work the setup should be exactly the same for each calibration to reduce errors and allow the operators to evolve a consistent rhythm for the comparison procedure.

The length of time needed for blocks to thermally equilibrate varies depending on block size and the contact area with the soaking plate. For blocks less than 10 mm long the same time can be used for all sizes. For longer sizes the contact area is important. If long blocks are laid down so their entire side is resting on the soaking plate then they will equilibrate much like the short blocks. If the blocks are wrapped in mylar this is no longer the case because the mylar and rubber bands used to hold the mylar on insulate the blocks from the plate. Wrapped blocks thus must stand upright on the plate. Since the thermal mass of the block increases but the contact area does not, the longer the block, the longer time it will take to come to thermal equilibrium. The minimum times used in our lab are shown in table 5.1

Table 5.1

Minimum Thermal Soaking Times in Minutes

Length (metric)	Length (English)	Time (min)
0 - 10 mm	0 - 0.25"	30
10 - 25 mm	0.3 - 1.0"	60
25 - 100 mm	2.0 - 4.0"	100
125 - 500 mm	5.0 - 20"	8 h

Experiments establishing optimum equalization times should be conducted in your own laboratory because of the many variables involved and differing measurement uncertainty requirements.

5.5.1 Handling Technique

The success of any intercomparison scheme is largely dependent on block handling techniques. These techniques include the insertion of all blocks between the styli in a like manner. The operator should develop a rhythm that, after practice, will ensure that each block is handled for approximately the same length of time. The time to make 24 observations should be approximately 2 to 4 minutes for an experienced operator with automatic data recording or an assistant doing the recording.

A camel hair brush or an air bulb is useful for sweeping or blowing dust particles from the blocks and the anvil just before insertion.

Short blocks are moved about by grasping them with rubber tipped tweezers or tongs. For handling square style blocks, the tips of the tweezers may be bent to accommodate this configuration.

For added insulation, light cotton gloves may be worn in addition to using tweezers when handling blocks from 10 mm to 100 mm (0.4 inch through 4 inches). Blocks above the 100 mm size are compared on a different instrument situated in a more precisely controlled temperature environment and the operator may wear a Mylar reflective cape as well as heavy gloves. A special pair of tongs is used to facilitate moving long blocks about the anvil and between the measuring probes. The design used is shown in figure 5.5. Operators have found that the least awkward place to grasp long blocks is about 10 mm above the bottom.

For our long block comparator the anvil stop (see figure 5.3) plays an important part in seating the probe on the block surface as well as positioning the gauge point between the probes. The block is

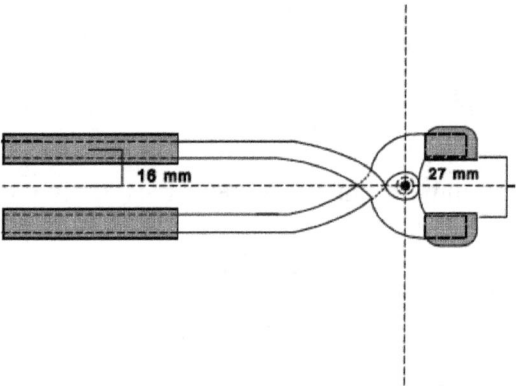

Figure 5.5. Tongs for moving long square cross section gauge blocks.

moved tangentially toward the stop allowing one corner to touch the stop and then with an angular motion (which seats the probe on the block surface) proceed to abut the entire edge of the block to the stop.

5.6 Comparison Designs

5.6.1 Drift Eliminating Designs

The generation and analysis of drift eliminating designs is described in appendix A. In this section we will discuss the four designs used for gauge block calibrations at NIST. The normal calibration scheme is the 12/4 design that uses two master blocks and two customer blocks. The design consists of all 12 possible comparisons between the four blocks. When only one set of customer blocks is available we use the 6/3 design which uses two master blocks and one customer block. The design consists of all six possible comparisons between the three blocks.

There are also two special designs used only for long gauge blocks. For long square cross section blocks (Hoke blocks) we use the 8/4 design which uses 8 comparisons among four blocks; two master blocks and two customer blocks. Long rectangular blocks are difficult to manipulate, particularly when over 200 mm. For these blocks we use a simplified measurement scheme using a master, a check standard, and one customer block in 4 measurements. For historical reasons this scheme is called "ABBA."

5.6.1.1 The 12/4 Design

Gauge blocks were calibrated using the 8/4 design for nearly 15 years. It was replaced in 1989 with a new design, the 12/4 design. The new scheme using 12 intercomparisons replaced the earlier scheme which used 8 intercomparisons. Increasing from 8 to 12 intercomparisons evolved from a determination to eliminate the systematic error caused by the varying contact deformation noted earlier. This design was described in detail in section 4.4.2.1. Here we will only discuss the reason for its development.

Previously, the restraint was the sum of the known lengths of block S and C, (S+C). The benefit of this restraint was that the percentage uncertainty is proportional to the uncertainties of the two blocks added together in quadrature. Since the uncertainty in S and C were generally about the same, the uncertainty in the restraint was reduced by a factor of 2. The drawback was that if the S, C, and customer blocks were not all the same material, corrections for the difference in stylus penetration was required. As discussed in the previous section this problem was found to result in measurable systematic errors. A study of our calibration history showed that contact pressure variations and the changing stylus geometry due to wear could not be controlled well enough to prevent nearly 25 nm of additional calibration uncertainty.

To reduce this problem one set of steel masters was replaced with chrome carbide masters. A new scheme was then designed using only the steel masters as the restraint for steel customer blocks, and the carbide masters as the restraint for carbide customer blocks. This eliminates the penetration correction when steel or chrome carbide blocks are calibrated, but it also increases the uncertainty in the restraint. To lower the total uncertainty the number of measurements was increased, offsetting the increased uncertainty in the restraint by reducing the comparison process random error.

5.6.1.2 The 6/3 Design

On some occasions two sets with the same sizes are not available for calibration. In the past the one set was used as both X and Y, giving two answers for each block, the average of the two being used for the report. To eliminate excess measurements a new design for one customer block was developed. The six comparisons are given below:

$$
\begin{aligned}
y_1 &= S - C \\
y_2 &= X - S \\
y_3 &= C - X \\
y_4 &= C - S \\
y_5 &= X - C \\
y_6 &= S - X
\end{aligned}
\tag{5.20}
$$

When the length L is the master block, the lengths of the other blocks are given by the following equations. The parameter Δ is the length drift/(time of one comparison).

$$S = L \tag{5.21a}$$

$$C = (1/6)(-2y_1 + y_2 + y_3 + 2y_4 - y_5 - y_6) + L \tag{5.21b}$$

$$X = (1/6)(-y_1 + 2y_2 - y_3 + y_4 + y_5 - 2y_6) + L \tag{5.21c}$$

$$<\Delta> = (1/6)(-y_1 - y_2 - y_3 - y_4 - y_5 - y_6) \tag{5.21d}$$

Formulas for the standard deviation for these quantities are in appendix A.

5.6.1.3 The 8/4 Design

The original 8/4 comparison design is currently only used for long rectangular blocks (100 mm and longer). This original design was drift eliminating but one comparison was used twice and there were several consecutive measurements of the same block. In the current (and new) 8/4 scheme, like the new 12/4 design, the length of only one block is used as the restraint, resulting in a slightly larger uncertainty. For long blocks the 12/4 design is not appropriate because the added time to make the extra comparisons allows drift to become less linear, an effect which increases the apparent random component of uncertainty.

The current 8/4 design is as follows:

$$
\begin{aligned}
y_1 &= S - C & y_2 &= X - Y \\
y_3 &= Y - S & y_4 &= C - X \\
y_5 &= C - S & y_6 &= Y - X \\
y_7 &= S - Y & y_8 &= X - C
\end{aligned}
\qquad (5.22)
$$

For the case where S is the master block (known length) the values of the other blocks are:

$$S = L \qquad (5.23a)$$

$$C = (1/12)(-4y_1 - 2y_3 + 2y_4 + 4y_5 - 2y_7 + 2y_8) + L \qquad (5.23b)$$

$$X = (1/12)(-2y_1 + 2y_2 + 3y_3 + 3y_4 + 2y_5 - 5y_7 - y_8) + L \qquad (5.23c)$$

$$Y = (1/12)(-2y_1 - 2y_2 + y_3 + 5y_4 + 2y_6 - 3y_7 - 3y_8) + L \qquad (5.23d)$$

$$\Delta = (-1/8)(y_1 + y_2 + y_3 + y_4 + y_5 + y_6 + y_7 + y_8) \qquad (5.23e)$$

5.6.1.4 The ABBA Design

The actual measurement scheme for this design is X-S-C-X, where X is the customer block, S and C are the NIST standard and control blocks. The name ABBA is somewhat of a misnomer, but has been attached to this design for long enough to have historical precedence.

The design is used only for long (over 200 mm or 8 in) rectangular blocks. These blocks have a very severe aspect ratio (height/width) and are very difficult to slide across the platen without risk of tipping. The X block is set on the anvil and under the measurement stylus when the blocks are set up the evening before they are measured. The S and C master blocks are also arranged on the anvil. The next morning the X block need only be move twice during the measurement, reducing the possibility of tipping. The measurement is repeated later and the two answers averaged. Obviously the F-test cannot be used, but the value of S-C is used as the control in a manner similar to the other measurement schemes.

5.6.2 Example of Calibration Output Using the 12/4 Design

```
1      GAUGE PROGRAM        Tuesday            September 10, 1991         11:01:28
2
3      Calibration:       4.000000 Inch Gauge Blocks                    Observer: TS
4      Block Id's:    S: 4361    C: 1UNG    X: XXXXXXX    Y: YYYYY      Federal: F4
5
6                               Observations              Observations
7      Blocks    at Ambient Conditions    at Standard Conditions    Differences    Residuals
8      ---------------------------------------------------------------------------------------
9      S ** C   52.88    62.51              66.28    72.92          -6.64          -1.00
10     Y ** S   62.50    53.32              72.91    66.71           6.20            .22
11     X ** Y   63.82    62.28              77.21    72.68           4.52           1.07
12     C ** S   62.18             54.15     72.59    67.54           5.05           -.58
13     C ** X   62.19    63.36              72.60    76.75          -4.15           -.35
14     Y ** C   63.15    62.26              73.55    72.66            .89            .55
15     S ** X   53.87    62.69              67.26    76.09          -8.80            .62
16     C ** Y   62.06    62.19              72.46    72.60           -.14            .22
17     S ** Y   53.57    62.87              66.96    73.27          -6.32           -.33
18     X ** C   62.76    62.21              76.15    72.62           3.53           -.27
19     X ** S   62.86    53.77              76.25    67.16           9.09           -.34
20     Y ** X   62.49    62.77              72.90    76.17          -3.27            .19
21
22     Obs Within S.D.=      .68         Obs Control=     -5.64        F-test =  6.89
23     Acc Within S.D.=      .26         Acc Control=     -8.90        T-test =  9.07
24     Acc Group S.D.=       .36         Temperature=     19.71
25
26                    Deviation
27     Serial      Nominal Size     from Nominal     Total Uncertainty    Coef.     Block
28     Number      (inches)         (microinch)      (microinch)          ppm/C     Material
29     ------      ------------     ------------     -----------------    -----     --------
30     4361        4.000000         -8.32            1.86                 11.50     steel
31     1UN6        4.000000         -2.68            1.86                  8.40     chrome carbide
32     MARTMA      4.000000         -1.86            1.89                 11.50     steel
33     TELEDY      4.000000          0.65            2.53                  8.40     chrome carbide
34
35     Process not in statistical control ...
```

Comments referenced by line numbers:
----

1. The header shows which computer program was used (GAGE), the date and time of the calibration. The date is recorded in the MAP file.

3. Block size and operator are identified and later recorded in the MAP file.

4. Block IDs are given here. The S and C blocks are NIST master blocks. The X and Y blocks are shortened forms of customer company names, represented here by strings of Xs and Ys. The ID of the comparator used in the calibration, F4, is also given and recorded in

the MAP file.

9-20. These lines provide the comparison data at ambient temperature and corrected to 20 °C using the thermal expansion coefficients in lines 30-34, and the temperature recorded in line 24. The differences are those between the corrected data. In the final column are the residuals, i.e., the difference between the best fit to the calibration data and the actual measured values.

22-25 These lines present statistical control data and test results.

The first column of numbers shows observed and accepted short term (within) standard deviation. The ratio of (observed/accepted) squared is the F-test value, and is shown in column 3. The last number, group S.D. is the long term standard deviation derived from (S-C) data in the MAP file.

The top line in the second column shows the observed difference (S-C), between the two NIST masters. The second line shows the accepted value derived from our history data in the MAP file. The difference between these two numbers is compared to the long term accepted standard deviation (group S.D. in column 1) by means of the t-test. The ratio of the difference and the group S.D. is the value of the t-test shown in column 3.

30-33 These lines present the calibration result, gauge block material and thermal expansion coefficient used to correct the raw data. Note that in this test the two customer blocks are of different materials. The same data is used to obtain the results, but the length of the steel customer block is derived from the NIST steel master and the length of the chrome carbide customer block is derived from the NIST chrome carbide master.

35 Since the F-test and t-test are far beyond the SPC limits of 2.62 and 2.5 respectively, the calibration is failed. The results are not recorded in the customer file, but the calibration is recorded in the MAP file.

The software system allows two options following a failed test: repeating the test or passing on to the next size. For short blocks the normal procedure is to reclean the blocks and repeat the test because a failure is usually due to a dirty block or operator error. For blocks over 25 mm a failure is usually related to temperature problems. In this case the blocks are usually placed back on the thermal soaking plate and the next size is measured. The test is repeated after a suitable time for the blocks to become thermally stable.

5.7 Current System Performance

For each calibration the data needed for our process control is sent to the MAP file. The information recorded is:

1. Block size
2. Date
3. Operator
4. Comparator
5. Flag for passing or failing the F-test, t-test, or both
6. Value of (S-C) from the calibration
7. Value of σ_w from the calibration

Process parameters for short term random error are derived from this information, as discussed in chapter 4 (see sections 4.4.2.1 and 4.4.2.2). The (S-C) data are fit to a straight line and deviations from this fit are used to find the standard deviation σ_{tot} (see 4.4.3). This is taken as an estimate of long term process variability. Recorded values of σ_w are averaged and taken as the estimate of short term process variability. Except for long blocks, these estimates are then pooled into groups of about 20 similar sizes to give test parameters for the F-test and t-test, and to calculate the uncertainty reported to the customer.

Current values for these process parameters are shown in figure 5.6 (σ_{tot}) and table 5.2. In general the short term standard deviation has a weak dependence on block length, but long blocks show a more interesting behavior.

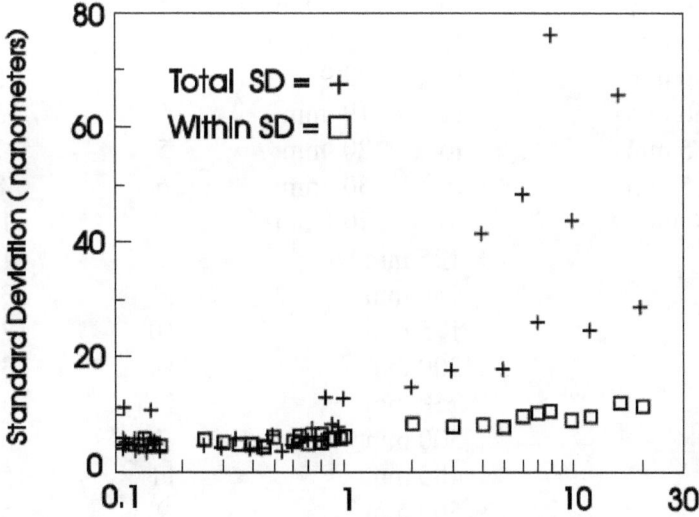

Figure 5.6. Dependence of short term standard deviation, σ_w, and long term standard deviation, σ_{tot}, on gauge block length.

Table 5.2

Table of σ_{within} and σ_{total} by groups, in nanometers

Group				σ_{within}	σ_{total}
2	0.100 in.	to	0.107 in.	5	4
3	0.108 in.	to	0.126 in.	5	6
4	0.127 in.	to	0.146 in.	5	5
5	0.147 in.	to	0.500 in.	5	5
6	0.550 in.	to	2.00 in.	6	8
7			3 in.	11	18
			4 in.	15	41
			5 in.	10	18
			6 in.	12	48
			7 in.	11	26
			8 in.	10	76
			10 in.	10	43
			12 in.	10	25
			16 in.	17	66
			20 in.	13	29

Group				σ_{within}	σ_{total}
14	1.00 mm	to	1.09 mm	5	6
15	1.10 mm	to	1.29 mm	5	5
16	1.30 mm	to	1.49 mm	5	5
17	1.50 mm	to	2.09 mm	5	5
18	2.10 mm	to	2.29 mm	5	5
19	2.30 mm	to	2.49 mm	5	5
20	2.50 mm	to	10 mm	5	5
21	10.5 mm	to	20 mm	5	7
22	20.5 mm	to	50 mm	6	8
23	60 mm	to	100 mm	8	18
24			125 mm	9	19
			150 mm	11	38
			175 mm	10	22
			200 mm	14	37
			250 mm	11	48
			300 mm	7	64
			400 mm	11	58
			500 mm	9	56

The apparent lack of a strong length dependence in short term variability, as measured by σ_w, is the result of extra precautions taken to protect the thermal integrity of longer blocks. Since the major cause of variation for longer blocks is thermal instability the precautions effectively reduce the length dependence.

One notable feature of the above table is that short term variability measured as σ_w and long term variability, measured as σ_{tot}, are identical within the measurement uncertainties until the size gets larger than 50 mm (2 in). For short sizes, this implies that long term variations due to comparators, environmental control and operator variability are very small.

A large difference between long and short term variability, as exhibited by the long sizes, can be taken as a signal that there are unresolved systematic differences in the state of the equipment or the skill levels of the operators. Our software system records the identity of both the operator and the comparator for each calibration for use in analyzing such problems. We find that the differences between operators and instruments are negligible. We are left with long term variations of the thermal state of the blocks as the cause of the larger long term variability. Since our current level of accuracy appears adequate at this time and our thermal preparations, as described earlier, are already numerous and time consuming we have decided not to pursue these effects.

5.7.1 Summary

Transferring length from master blocks to customer blocks always involves an uncertainty which depends primarily on comparator repeatability and the number of comparisons, and the accuracy of the correction factors used.

The random component of uncertainty (σ_{tot}) ranges from 5 nm (0.2 µin) for blocks under 25 mm to about 75 nm (3 µin) for 500 mm blocks. This uncertainty could be reduced by adding more comparisons, but we have decided that the gain would be economically unjustified at this time.

Under our current practices no correction factors are needed for steel and chrome carbide blocks. For other materials a small added uncertainty based on our experience with correction factors is used. At this time the only materials other than steel and chrome carbide which we calibrate are tungsten carbide, chrome plated steel and ceramic, and these occur in very small numbers.

6. Gauge Block Interferometry

6.1 Introduction

Gauge Block calibration at NIST depends on interferometric measurements where the unit of length is transferred from its definition in terms of the speed of light to a set of master gauge blocks which are then used for the intercomparison process.

Gauge blocks have been measured by interferometry for nearly 100 years and the only major change in gauge block interferometry since the 1950's has been the development of the stabilized laser as a light source. Our measurement process employs standard techniques of gauge block interferometry coupled with an analysis program designed to reveal random and systematic errors. The analysis program fosters refinements aimed at reducing these errors. A practical limit is eventually reached in making refinements, but the analysis program is continued as a measurement assurance program to monitor measurement process reliability.

Briefly, static interferometry is employed to compare NIST master gauge blocks with a calibrated, stabilized laser wavelength. The blocks are wrung to a platen and mounted in an interferometer maintained in a temperature controlled environment. The fringe pattern is photographed and at the same moment those ambient conditions are measured which influence block length and wavelength. A block length computed from these data together with the date of measurement is a single record in the history file for the block. Analysis of this history file provides an estimate of process precision (long term repeatability), a rate of change of length with time, and an accepted value for the block length at any given time.

Gauge block length in this measurement process is defined as the perpendicular distance from a gauging point on the top face of the block to a plane surface (platen) of identical material and finish wrung to the bottom face. This definition has two advantages. First, specifying a platen identical to the block in material and surface finish minimizes phase shift effects that may occur in interferometry. Second, it duplicates practical use where blocks of identical material and finish (from the same set) are wrung together to produce a desired length. The defined length of each block includes a wringing layer, eliminating the need for a wringing layer correction when blocks are wrung together.

The NIST master gauge blocks are commercially produced gauge blocks and possess no unusual qualities except that they have a history of calibration from frequent and systematic comparisons with wavelengths of light.

6.2 Interferometers

Two types of interferometers are used at NIST for gauge block calibration. The oldest is a Kösters type interferometer, and the newest, an NPL Gauge Block Interferometer. They date from the late 1930's and 1950's respectively and are no longer made. Both were designed for multiple wavelength interferometry and are much more complicated than is necessary for single wavelength laser interferometry. The differences are pointed out as we discuss the geometry and operation of both interferometers.

6.2.1 The Kösters Type Interferometer

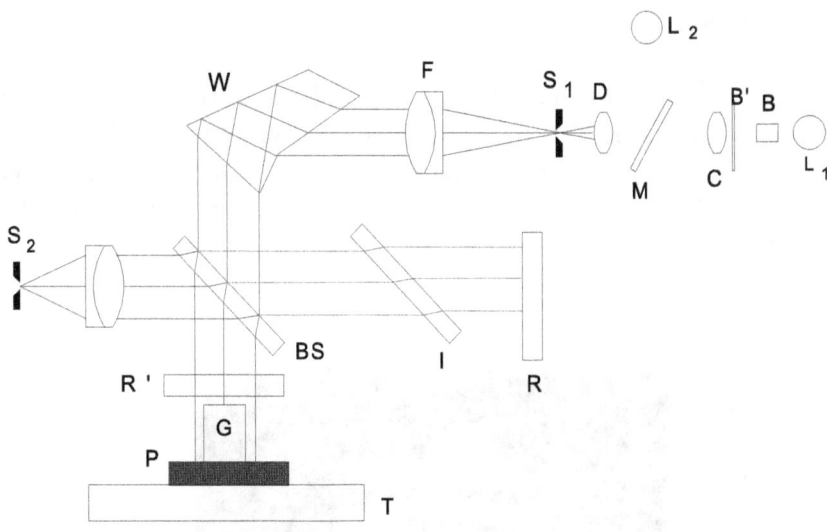

Figure 6.1 Kösters type gauge block interferometer.

The light beam from laser L1 passes through polarization isolator B and spatial filter B', and is diverged by lens C. This divergence lens is needed because the interferometer was designed to use an extended light source (a single element discharge tube). The laser produces a well collimated beam which is very narrow, about 1 mm. The spatial filter and lens diverges the beam to the proper diameter so that lens D can focus the beam on the entrance slit in the same manner as the light from a discharge tube. The entrance slit (S_1) is located at the principle focus of lenses D and F. Lens F collimates the expanded beam at a diameter of about 35 mm, large enough to encompass a gauge block and an area of the platen around it.

The extended collimated beam then enters a rotatable dispersion prism, W. This prism allows one wavelength to be selected from the beam, refracting the other colors at angles that will not produce interference fringes. An atomic source, such as cadmium ($_L$) and a removable mirror M, can be used in place of the laser as a source of 4 to 6 different calibrated wavelengths. If only a laser is used the prism obviously has no function, and it could be replaced with a mirror.

The rest of the optics is a standard Michelson interferometer. The compensator plate, I, is necessary when using atomic sources because the coherence length is generally only a few centimeters and the effective light path in the reference and measurement arms must be nearly the same to see fringes.

Any helium-neon laser will have a coherence length of many meters, so when a laser is used the compensator plate is not necessary.

The light beam is divided at the beam splitter top surface, BS, into two beams of nearly the same intensity. One beam (the measuring beam) continues through to the gauge block surface G and the platen surface P. The other beam (reference beam) is directed through the compensating plate I to the plane mirror R. The beams are reflected by the surfaces of the mirror, platen, and gauge block, recombined at the beam splitter, and focused at the exit aperture, S_2. A camera or human eye can view the fringes through the exit aperture.

Interference in this instrument is most readily explained by assuming that an image of reference mirror R is formed at R' by the beam splitter. A small angle, controlled by adjustment screws on table T, between the image and the gauge block platen combination creates a Fizeau fringe pattern. When this wedge angle is properly adjusted for reading a gauge block length, the interference pattern will appear as in figure **6.2**.

Fig. 6.2 Observed fringe pattern for a gauge block wrung to a platen.

The table T has its center offset from the optical axis and is rotatable by remote control. Several blocks can be wrung to individual small platens or a single large platen and placed on the table to be moved, one at a time, into the interferometer axis for measurement.

Two criteria must be met to minimize systematic errors originating in the interferometer: (1) the optical components must be precisely aligned and rigidly mounted and (2) the optical components must be of high quality, i.e., the reference mirror, beam splitter and compensator must be plane and the lenses free of aberrations.

Alignment is attained when the entrance aperture, exit aperture and laser beam are on a common axis normal to, and centered on, the reference mirror. This is accomplished with a Gaussian eyepiece, having vertical and horizontal center-intersecting cross-hairs, temporarily mounted in place of the exit aperture, and vertical and horizontal center-intersecting cross-hairs permanently mounted on the reference mirror. Through a process of autocollimation with illumination at both entrance aperture and Gaussian eyepiece, the reference mirror is set perpendicular to an axis through the intersections of the two sets of cross-hairs and the entrance, and exit apertures are set coincident with this axis.

In addition, the laser beam is aligned coincident with the axis and the prism adjusted so the laser light coming through the entrance aperture is aligned precisely with the exit aperture. Having an exact 90° angle between the measuring leg and the reference leg is not essential as can be seen from instrument geometry. This angle is governed by the beam splitter mounting. All adjustments are checked regularly.

The gauge block and its platen are easily aligned by autocollimation at the time of measurement and no fringes are seen until this is done. Final adjustment is made while observing the fringe pattern, producing the configuration of figure 6.2.

Temperature stability of the interferometer and especially of the gauge block being measured is important to precise measurement. For this reason an insulated housing encloses the entire Kosters interferometer to reduce the effects of normal cyclic laboratory air temperature changes, radiated heat from associated equipment, operator, and other sources. A box of 2 cm plywood with a hinged access door forms a rigid structure which is lined with 2.5 cm thick foam plastic. Reflective aluminum foil covers the housing both inside and out.

6.2.2 The NPL Interferometer

The NPL interferometer is shown schematically in figure 6.3. The system is similar in principle to the Kösters type interferometer, although the geometry is rather different.

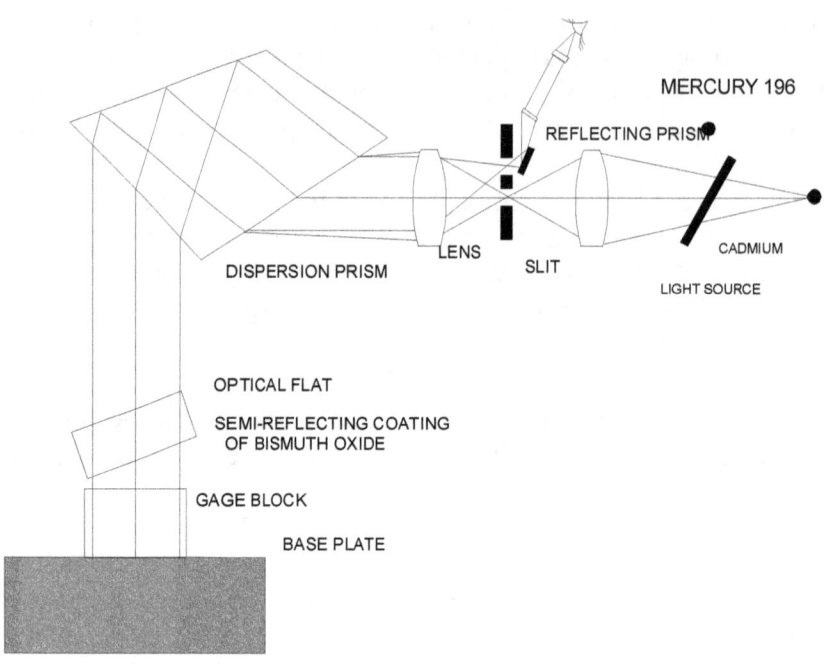

Figure 6.3 Schematic of the NPL interferometer.

The NIST version differs slightly from the original interferometer in the addition of a stabilized laser, L_1. The laser beam is sent through a spatial filter B (consisting of a rotating ground glass plate) to destroy the temporal coherence of the beam and hence reduce the laser speckle. It is then sent through a diverging lens C to emulate the divergent nature of the atomic source L_2. The atomic source, usually cadmium, is used for multicolor interferometry.

The beam is focused onto slit, S_1, which is at the focal plane of the D. Note that the entrance slit and exit slit are separated, and neither is at the focal point of the lens. Thus, the light path is not perpendicular to the platen. This is the major practical difference between the Kösters and NPL interferometers. The Kösters interferometer slit is on the optical axis that eliminates the need for an obliquity correction discussed in the next section. The NPL interferometer, with the slit off the optical axis does have an obliquity correction. By clever design of the slits and reflecting prism the obliquity is small, only a few parts in 10^6.

The beam diverges after passing through slit S_1 and is collimated by lens F. It then passes through the wavelength selector (prism) W and down onto the Fizeau type interferometer formed by flat R and platen P. The height of the flat can be adjusted to accommodate blocks up to 100 mm. The adjustable mount that holds the flat can be tilted to obtain proper fringe spacing and orientation shown in figure 6.2.

The light passes back through the prism to a second slit S2, behind which is a small reflecting prism (RP) to redirect the light to an eyepiece for viewing.

The optical path beyond the slit is inside a metal enclosure with a large door to allow a gauge block platen with gauge blocks wrung down to be inserted. There are no heat sources inside the enclosure and all adjustments can be made with knobs outside the enclosure. This allows a fairly homogeneous thermal environment for the platen and blocks.

Multicolor interferometry is seldom needed, even for customer blocks. A single wavelength is adequate if the gauge block length is known better than 1/2 fringe, about 0.15 μm (6 μin). For customer calibrations, blocks are first measured by mechanical comparison. The length of the block is then known to better than 0.05 μm (2 μin), much better than is necessary for single wavelength interferometry.

Since the NPL interferometer is limited to relatively short blocks the need for thermal isolation is reduced. The temperature is monitored by a thermistor attached to the gauge block platen. One thermometer has proven adequate and no additional insulation has been needed.

6.2.3 Testing Optical Quality of Interferometers

Distortion caused by the system optics, is tested by evaluating the fringe pattern produced on a master optical flat mounted on the gauge block platen support plate. Photographs showing fringes oriented vertically and horizontally are shown in figure 6.4. Measurements of the fringe pattern indicate a total distortion of 0.1 fringe in the field. A correction could be applied to each block depending on the section of the field used, but it would be relatively small because the field section used in fringe fraction measurement is small. Generally the corrections for optical distortion in both instruments are too small to matter.

Figure 6.4 The quality of the optics of the interferometer can be measured by observations of the fringes formed by a calibrated reference surface such as an optical flat or gauge block platen.

6.2.4 Interferometer Corrections

In some interferometers, the entrance and exit apertures are off the optical axis so light falls obliquely on the gauge block. The NPL interferometer is one such design.

In the Kosters type interferometer these apertures are aligned on the optical axis and consequently the illumination is perpendicular to the gauge block and platen surfaces. In figure 6.5 the case for normal incidence is given. Note that the light reflected from the surfaces follows the same path before and after reflection. At each point the intensity of the light depends on the difference in phase between the light reflected from the top surface and that reflected from the lower surface. Whenever this phase difference is 180° there will be destructive interference between the light from the two surfaces and a dark area will be seen. The total phase difference has two components. The first component results from the electromagnetic phase difference of reflections between the glass to air interface (top flat) and the air to glass interface (bottom flat). This phase shift is (for non-conducting platens) 180°. The second phase difference is caused by the extra distance traveled by the light reflected at the bottom surface. The linear distance is twice the distance between the surfaces, and the phase difference is this difference divided by the wavelength, λ.

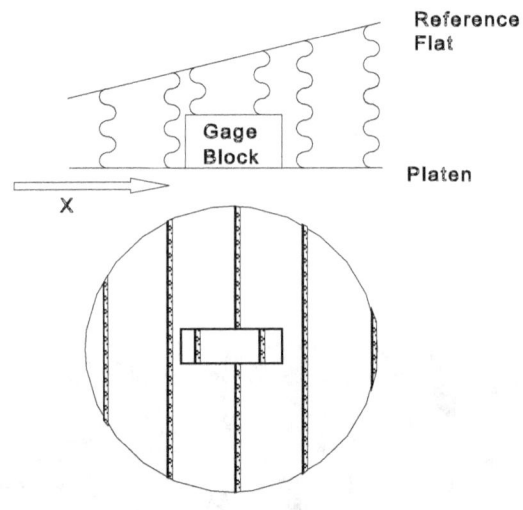

Figure 6.5 In the ideal case of normal incidence, a dark fringe will appear wherever the total path difference between the gauge block and platen surfaces is a multiple of the wavelength.

At every position of a dark fringe this total phase difference due to the path difference is a multiple of λ:

$$\Delta(\text{path}) = n\lambda = 2D \tag{6.1}$$

where n is an integer.

If the wedge angle is α, then the position on the nth fringe is

$$X = (n\lambda)/(2\tan(\alpha)) \qquad (6.2)$$

For the second case, where the light is not normal to the block and platen, we must examine the extra distance traveled by the light reflecting from the lower flat and apply a compensating factor called an obliquity correction. Figure 6.6 shows this case. The difference in path between the light reflected from the upper reference surface and the lower reference surface must be a multiple of λ to provide a dark fringe. For the nth dark fringe the distance 2L must equal nλ.

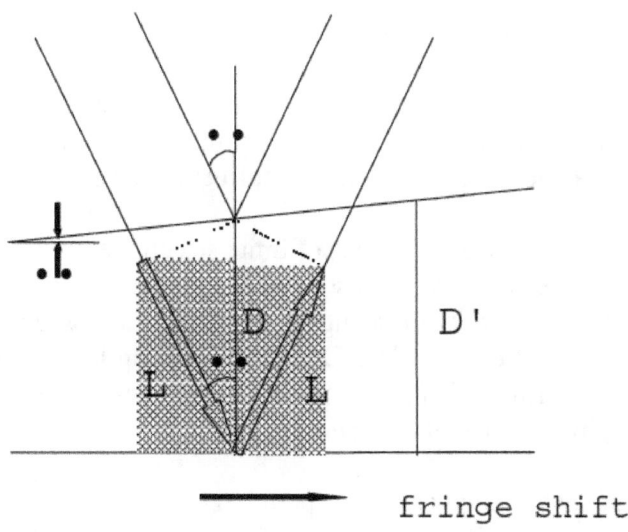

Figure 6.6 When the light is not normal to the surface the fringes are slightly displaced by an amount proportional to the distance between the surfaces.

The equation for a dark fringe then becomes

$$\Delta(\text{path}) = n\lambda = 2L \qquad (6.3)$$

From the diagram we see that L is:

$$L = D\cos(\theta) = n\lambda \qquad (6.4)$$

Thus the new perpendicular distance, D', between the two flats must be larger than D. This implies that the nth fringe moves away from the apex of the wedge, and the distance moved is proportional to cos(θ) and distance D.

When a gauge block is being measured, the distance from the top flat to the bottom flat is larger than the distance from the top flat to the gauge block surface. Thus the fringes on the block do not shift as far as the fringes on the flat. This larger shift for the reference fringes causes the fringe fraction to be reduced by a factor proportional the height of the block, H, and cos(θ). Note that this effect causes the block to appear shorter than it really is. Since the angle θ is small the cosine can be approximated by the first terms in its expansion

$$\cos(\theta) = 1 - \frac{\theta^2}{2} + \frac{\theta^4}{24} + \ldots \quad (6.5)$$

and the correction becomes:

$$\text{Obliquity Correction} \sim (H/2)\,\theta^2 \quad (6.6)$$

This correction is important for the NPL Hilger interferometer where the entrance slit and observation slit are separated. In the Kösters type interferometer the apertures are precisely on the optical axis, the obliquity angle is zero, and thus no correction is needed.

In most interferometers, the entrance aperture is of finite size, therefore ideal collimation does not occur. Light not exactly on the optical axis causes small obliquity effects proportional to the size of the aperture area. A laser, as used with these interferometers, is almost a point source because the aperture is at the common focal point of lens D, and collimating lens E. At this point the beam diameter is the effective aperture. When using diffuse sources, such as discharge lamps, the effective aperture is the physical size of the aperture.

The theory of corrections for common entrance aperture geometries has been worked out in detail [37,38,39]. The correction factor for a gauge block of length L, measured with wavelength λ, using a circular aperture is approximately [40]

$$\text{Slit Correction (Circle)} = \frac{L x D^2}{16 f^2} \quad (6.7)$$

where L is the block length, D is the aperture diameter, and f is the focal length of the collimating lens. For rectangular apertures the approximate correction is

$$\text{Slit Correction (Rectangle)} = \frac{L x (h^2 + l^2)}{24 f^2} \quad (6.8)$$

where L is the block length, h and l the height and length of the slit, and f the focal length of the collimating lens.

As aperture diameter approaches zero, the correction approaches zero. The correction is very small when using a laser because the beam diameter is small at the aperture.

6.2.5 Laser Light Sources

The advantage of a laser light source lies in its unequalled coherence. Conventional spectral light sources have such low coherence that blocks longer than 300 mm have, in the past, been measured by stepping up from shorter blocks. Laser fringe patterns in these interferometers compared with patterns from other spectral lamps are of superior contrast and definition through the length range 0 to 500 mm.

The disadvantage stems from the somewhat unstable wavelength of lasers. This problem has been overcome by a number of different stabilization schemes, and by using the .6328 µm line of HeNe lasers. These lasers can be related to the definition of the meter by calibration against an iodine stabilized HeNe laser [41,42]. The wavelength of the iodine stabilized laser is known to a few parts in 10^9 and is one of the recommended sources for realizing the meter [8]. The lasers currently used for gauge block calibration are polarization stabilized [43].

Single wavelength static interferometry, in contrast with multiple wavelength interferometry, requires that the gauge block length be known to within 0.25 wavelength (0.5 fringe) either from its history or from another measurement process. This is no problem for NIST reference blocks, or for blocks calibrated by the NIST mechanical comparison technique.

Laser light must not be allowed to reflect back into its own cavity from the interferometer optical components because this will disturb the lasing action and may cause a wavelength shift, wavelength jittering or it may stop the lasing altogether. The reflected light is prevented from re-entering the laser by a polarization isolator consisting of a Glan Thompson prism and a quarter wave plate in the beam as it emerges from the cavity. This assembly is tilted just enough to deflect reflections from its faces to the side of the laser exit aperture.

6.3 Environmental Conditions and their Measurement

Environmental control of the laboratory holds temperature at 20.05 °C and water vapor content below 50% relative humidity. Temperature variations within the insulated interferometer housing are attenuated by a factor of about 10 from those in the room, thus insuring stability of both interferometer and blocks.

Temperature, atmospheric pressure, and water vapor content of the ambient air in the interferometer light path must be measured at the time the gauge block is measured. From these properties the refractive index of the air is calculated, and in turn, the laser wavelength. Since interferometry is a comparison of the light wavelength to the block length the accuracy of the wavelength is of primary importance. The gauge block temperature is measured so that the block length at 20 °C, the standard temperature for dimensional metrology, can be computed.

6.3.1 Temperature

The measuring system accuracy for air and block temperature depends on the length of the gauge block to be measured. Since both the wavelength and block temperature corrections are length dependent, a 500 mm measurement must have correction factors 10 times as good as those for a 50 mm measurement to achieve the same calibration accuracy.

For blocks measured with the NPL interferometer (less than 100 mm) a calibrated thermistor probe is used to measure temperature to an accuracy of 0.01 °C and a calibrated aneroid barometer is used to measure atmospheric pressure to 0.1 mm of mercury.

For long blocks the system is much more complicated. The measuring system for air and block temperature consists of copper-constantan thermocouples referenced to a platinum resistance thermometer. Figure 6.7 is a schematic diagram of the system.

A solid copper cube provides a stable reference temperature measured by a Standard Platinum Resistance Thermometer (SPRT) and a Mueller resistance bridge. A well in the cube holds the thermometer stem, the reference thermocouple junction, and a liquid of high heat conductivity but low electrical conductivity. There is also a second well whose function will be described later. An enclosed thermocouple selector switch connects the measuring junctions, one at a time, to the reference junction and the thermal emfs generated by the temperature differences between measuring junction and reference junction are indicated on a nanovoltmeter. The nanovoltmeter is shared with the bridge where it is used as a nullmeter. To keep the copper block in a stable state and to minimize thermal emfs in the selector switch, both block and switch are enclosed in an insulated box. The switch is operated through a fiber shaft extending to the exterior and the protruding part of the thermometer is protected by an insulated tube.

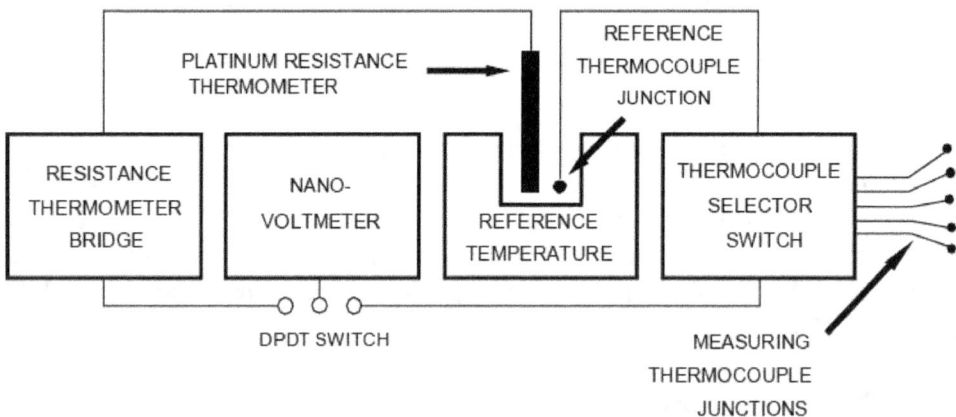

Figure 6.7 The temperature measurement system used for gauge block calibration consists of multiple thermocouples referenced to a calibrated platinum resistance thermometer.

Thermocouples generate emfs proportional to the temperature gradient along the wire:

$$E = \int_{T_2}^{T_1} \phi \, dT \tag{6.9}$$

where T_1 and T_2 are the temperatures of the reference and measuring junctions respectively, and ϕ is the thermocouple constant. Minimizing this gradient reduces uncertainties.

Relating the system to the International Temperature Scale of 1990 (ITS '90) is a three step procedure. First, the SPRT is calibrated using methods described in NIST Technical Note 1265 [44]. Second, the bridge is calibrated as described the same monograph. The third step is the calibration of the thermocouples, which is accomplished with a second SPRT and insulated copper cube. The second cube has an extra well for the thermocouples being calibrated and is heated one or two degrees to equilibrium. Measured thermal emfs generated by the temperature difference between the two cubes together with the temperatures of the cubes allow computation of a calibration factor for each junction.

Integrity is maintained by periodic check of the SPRT with a triple point cell, checking the bridge against a calibrated resistor and checking the thermocouples for equality when they are all at the same temperature in a well of an insulated copper cube.

6.3.2 Atmospheric Pressure

The interferometer housing is not airtight and pressure inside it is the same as laboratory pressure. An aneroid barometer is located adjacent to and at the same elevation as the interferometer. The aneroid used for gauge block measurements has good stability, and frequent comparisons with NIST reference barometers insure its reliability.

For long blocks more precise measurement of the pressure is useful. There are a number of commercially available atmospheric pressure sensors available which are of adequate accuracy for gauge block calibration. Remembering that measurements to one part in 10^7 requires pressure measurement accuracy of .3mm of Hg, a calibrated aneroid barometer can be adequate. For systems requiring automatic reading of pressure by a computer there are a number of barometers which use capacitance or vibrating cylinders to measure pressure and have RS-232 or IEEE 488 interfaces.

6.3.3 Water Vapor

The humidity is derived from a chilled mirror dew point hygrometer. The system is calibrated by the NIST humidity calibration service and is accurate to about 1% R.H. Other systems have adequate accuracy but need more frequent checking for drift. The disadvantage of the chilled mirror is that it requires a specified air flow rate not usually available inside an interferometer. Humidity inside an interferometer enclosure may differ from the laboratory value by as much as 6% R.H., especially during periods of changing laboratory humidity. If this presents a serious problem (see table 6.1) then a hygrometer type less dependent on air flow rate must be mounted inside the interferometer.

6.4 Gauge Block Measurement Procedure

In preparation for measurement, the gauging faces of the NIST standards are cleaned with ethyl alcohol. Wiping the alcohol-rinsed gauging faces is done with paper tissue or a clean cotton towel. Lint or dust is removed with a camel's hair brush.

The gauging faces are examined for burrs and, if present, these are removed with a deburring stone. The bottom end is tested for wringability with a quartz optical flat. The transparency of the flat makes possible a judgment of the wring. A uniform grey color at the interface indicates a good wring. Any colored or light areas indicate a gap of more than 25 nm between the faces in that area, and such a gap may result in an erroneous length measurement. Deburring and cleaning is continued until a good wring is achieved. The quartz flat is left wrung to the gauging face to keep it clean until preparations are completed for wringing the block to the steel optical flat (platen).

Preparation of the steel platen is similar. Ethyl alcohol is used to rinse the face, then deburring if necessary, followed by more rinsing and drying. After sweeping with a camel hair brush, a thin uniform film of light grease is applied to the wringing surface with a bit of tissue. This film is rubbed and finally polished with tissue or towel until the film appears to be gone. After carefully sweeping the platen with a brush, it is ready for wringing.

The block is removed from the quartz flat and the exposed gauging face is immediately wrung by carefully sliding it onto the edge of the platen, maintaining perpendicularity by feel. Sliding is continued until wringing resistance is felt, and then with slight downward pressure the block is slowly worked into its final position. Square style blocks, such as the NIST long blocks, are positioned so that the gauge point is at the right in the viewing field of the interferometer. One to four blocks can be wrung to the platens used in the Kosters interferometer. Ten square blocks or 20 rectangular blocks can be wrung onto the platens at one time for use in the NPL interferometer.

For long blocks measured in the Kosters interferometer, the platen with its wrung blocks is placed in the interferometer and two thermocouples are fed down the center hole of each block, one about three quarters down and one about one quarter down the length of the block. A small wad of polyurethane is pushed into the hole to seal and hold the wires. For small blocks, where there can be no significant vertical thermal gradient, the temperature sensor is attached to the platen.

A preliminary alignment by autocollimation with the Gaussian eyepiece as the support table is adjusted, will produce fringes. Rapid changes in the fringe pattern occur as the blocks and platen cool from handling.

It is convenient to wring the blocks in late afternoon and allow overnight temperature stabilization. Two length observations are made several hours apart, but the first measurement is not taken until the laser is in equilibrium. Leaving the laser on continuously when observations are being made over several days eliminates delays.

Final fringe pattern adjustment is made so that the platen fringes are parallel to the top and bottom edges of the block and one fringe on the block goes through its defined gauge point. For the Kosters interferometer the direction of increasing fringe order is verified by pressing down on the eyepiece and observing the fringe movement. In the NPL, the reference mirror is pushed and the fringe movement is observed. The fringe fraction can be estimated by eye or a photograph can be taken of the pattern. Block temperature, air temperature, barometric pressure, and humidity are measured immediately before and after the observation and then averaged.

The photograph can be either a positive or a negative. A compromise is made between image size and exposure time, but short exposure is desirable because the fringe pattern may shift during a long exposure. Further image magnification takes place in the fringe measuring instrument. A compromise is also made between magnification in the photograph and magnification in the instrument used to measure the photograph. Too much total magnification degrades image sharpness, making fringe measurements more difficult.

Fringe fractions are obtained from the photograph with a coordinate comparator having a fixed stage in the focal plane of a microscope movable on X-Y slides by precision screws with drum readouts. Four measurements of distance "a" and "b" are recorded and averaged to obtain fringe fraction f, as shown in figure 6.8.

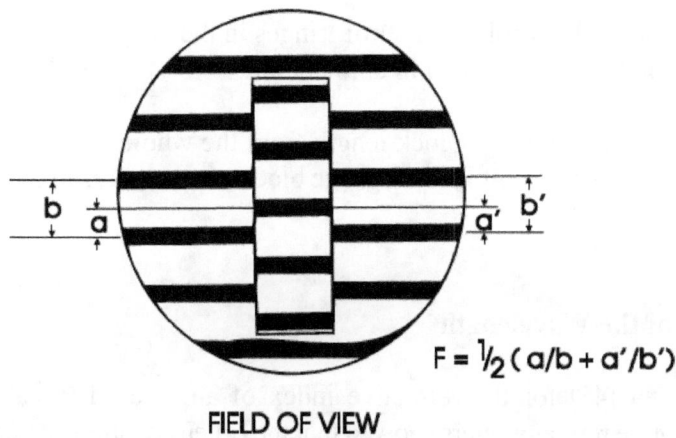

Figure 6.8. The fringe fraction is the fractional offset of the fringes on the block from those of the platen.

Settings on the block fringe to obtain "a" are made at the point where the fringe intersects the gauge point. Settings on the platen fringes to obtain "a" and "b" are made as close as practical to the edges of the block because it is here that the platen best represents an extension of the bottom face of the block. Parallelism between gauge block and platen is also read from the photo because this information is useful in detecting a defective wring. Poor wringing contact is indicated if the parallelism is different from its usual value (or mechanically determined value using the comparator) and this is sufficient cause to discard a particular length measurement.

Photographing fringe patterns has several advantages. The photograph is a permanent record that can be reinterpreted at any time with the same or other techniques. Changes in block geometry are readily seen by comparing photographs in time sequence. The coordinate comparator is superior to the common practice of fringe fraction estimation by eye because it is more objective, lends itself to multiple measurements and averaging, and is more precise. Finally, photography is fast and thus permits readings of the ambient conditions to closely bracket the fringe recording. This is especially important because atmospheric pressure is generally not constant.

The advent of computer vision systems provides another method of recording and analyzing fringes. There are commercially available software packages for measuring fringe fractions that can be incorporated into an automated gauge block measuring system.

6.5 Computation of Gauge Block Length

Calculating gauge block length from the data is done in 3 steps:

1. Calculation of wavelength, λ_{tpf}, at observed ambient conditions.

2. Calculation of the whole number of fringes in the gauge block length from its predicted length and the laser wavelength in ambient air.

3. Calculation of the gauge block length from the whole number of fringes, the observed fraction, wavelength in ambient air, gauge block temperature, and interferometric correction factors.

6.5.1 Calculation of the Wavelength

Edlen's 1966 formula [45] for the refractive index of air is used to calculate the wavelength. Comparison of absolute refractometers showed that the Edlen equation agreed with refractometers as well as the refractometers agreed with each other, to about 1 part in 10^7 [46]. More recent work by Birch and Downs [47] has shown that the original data used for the water vapor correction was slightly in error. The new correction (the parameter C below) is thought to increase the accuracy of the Edlen formula to about 3×10^{-8}. This latter figure can be taken as an estimate of the accuracy of the formula.

The latest revision of the Edlen formula is [48]:

$$\lambda_0 = n_{tpf} \lambda_{tpf} \qquad (6.10)$$

and thus

$$\lambda_{tpf} = \frac{\lambda_0}{n_{tpf}} = \lambda_0 [1 + A*B - C]^{-1} \qquad (6.11)$$

where

$$A = \frac{p[8342.22 + 2406057(130-\sigma^2)^{-1} + 15997(38.9-\sigma^2)^{-1}] 10^{-8}}{96095.43} \qquad (6.12)$$

$$B = \frac{1 + p(0.601-0.00972t) 10^{-6}}{1+0.003661t} \qquad (6.13)$$

$$C = f(3.7345-0.0401\sigma^2) 10^{-8}. \qquad (6.14)$$

The symbols used are:

λ_0 = Vacuum wavelength in μm
λ_{ptf} = Wavelength at observed conditions t,p,f
n_{ptf} = Index of refraction of air at t,p,f
p = Atmospheric pressure in Pa
t = Temperature in degrees Celsius
f = water vapor pressure in Pa
σ = $1/\lambda_0$ in μm.

6.5.2 Calculation of the Whole Number of Fringes

This method for determining the whole number of fringes is valid if the block length is known to better than 0.5 fringe (1 fringe is about 0.32 μm for helium neon laser light) from either its history or from an independent measurement process. The calculation is as follows:

$$F = \frac{2L_p[1+C(t'-20)]}{\lambda_{tpf}} \qquad (6.15)$$

where F is the number of interference fringes in the length of the block at temperature t', and in air of temperature t, pressure p, and vapor pressure f.

L_p is the predicted block length at 20 °C taken from its history or an independent measurement.

C is the linear thermal expansion coefficient per degree Celsius of the block.

t' is the block temperature at the time of measurement.

λ_{tpf} is the wavelength in the ambient air at the time of measurement.

The fractional part of F is retained temporarily for the reasons explained below.

6.5.3 Calculation of Block Length from the Observed Data

Generally, the observed fringe fraction φ is simply substituted for the fractional part of F, but there are cases where the last digit in whole number F must be raised or lowered by one. For example if the fractional part of F is 0.98 and the observed fraction is 0.03, obviously, the last whole number in F must be raised by one before substituting the observed fraction.

The new measured block value, in fringes, is

$$F' = F + \varphi. \tag{6.16}$$

Finally, the interferometer aperture correction, δ_2, is added, the block is normalized to 20 °C, and a conversion to length units is made to arrive at the final value at 20 °C as follows:

$$L_{20} = \frac{\lambda_{tpf}}{2} F'[1+C(20-t')] + \delta_2 . \tag{6.17}$$

6.6 Type A and B Errors

Potential sources of errors are as follows:

1. Wavelength
    a. Vacuum wavelength of laser
    b. Refractive index of air equation
    c. Refractive index determination
        Air temperature measurement
        Atmospheric pressure measurement
        Humidity measurement

2. Interferometer
    a. Alignment
    b. Aperture correction
    c. Obliquity correction

3. Gauge Block
    a. Gauge block temperature measurement
    b. Thermal expansion coefficient
    c. Phase shift difference between block and platen

Length proportional errors are potentially the most serious in long block measurement. Their relative influence is illustrated in table 10. One example will help explain the table: if air temperature measurement is in error by 0.1 °C a systematic error of 1 part in 10^7 results. For a 250 mm gauge block the error is 25 nm, a significant error. On a 10 mm gauge block the same temperature measurement leads to only a 1 nm error, which is negligible.

In our measurement process every practical means, as previously discussed, was used to reduce these errors to a minimum and, at worst, errors in measuring the parameters listed above do not exceed those listed in table 10. Further evaluation of these errors is described in section 4.3 on process control and our master block history.

A few sources of error in the list of systematic errors but not in the table have not been completely discussed. The first of these is phase shift difference between block and platen. When light is reflected from a surface the phase of the reflected light is shifted from the phase of the incident by some amount which is characteristic of the surface's material composition and surface finish. The explanation and measurement of phase is given in appendix C. Most gauge blocks have phase shifts that differ from quartz by an equivalent of 25 to 75 nm. This is a large source of error, although if both the block and platen are made of the same metal the error is reduced below 20 nm. Since the phase shift occurs only at the metal surface it is not length dependent.

Table 10

Effects of Environmental Factor Errors on Length

Parameter	Change Which Leads to a Length Error of 1 part in 10^7
Wavelength	
Vacuum wavelength	1 part in 10^7
Refractive index formula	1 part in 10^7
Air temperature	0.1°C
Atmospheric Pressure	40 pa
Relative Humidity	12% R.H.
Interferometer Alignment	0.45 mm/m
Gauge Block	
Steel gauge block temperature	0.009 °C
Chrome Carbide block temp.	0.012 degc

Thermal expansion coefficients of gauge blocks have an uncertainty that is, to a large extent, nullified by measuring at temperatures very close to the 20 °C reporting temperature. Gauge block manufacturers give 11.5×10^{-6}/°C for steel blocks. This average value may be uncertain by up to 0.6×10^{-6}/°C for individual blocks. The expansion coefficient for steel blocks is also length dependent for long blocks, as discussed earlier in section 3.3. The expansion coefficient for longer sizes can differ from the nominal 11.5×10^{-6}/°C by nearly 1×10^{-6}/°C. These uncertainties can cause significant errors if blocks are used at other than 20 °C.

There are, of course, random errors that arise from variability in any parameter in the table, and in addition, they can arise in the fringe fraction measurement and wringing variability. Wringing is discussed in more detail in appendix B. Gauge blocks are generally measured for 3 to 5 separate wrings to sample variability due to these sources.

6.7 Process Evaluation

Figure 6.9 shows some examples of the interferometric history of NIST master gauge blocks. As expected, the scatter in the data is larger for longer blocks because of the uncertainties in the length dependent thermal expansion and index of refraction corrections.

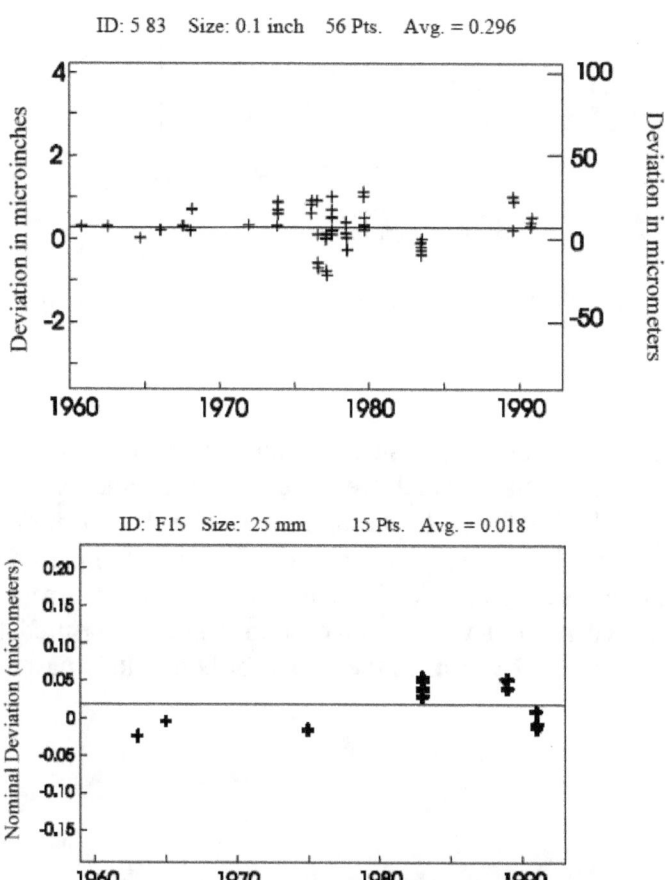

Figure 6.9.   Examples of the interferometric history of NIST master gauge blocks.

The length dependence of the variability is shown graphically in figure 6.10. This variability has two components: (1) non-length related variabilitys due to variations in the wringing film between the block and the platen, and random errors in reading the fringes; and (2) a length dependent part primarily due to environmental effects.

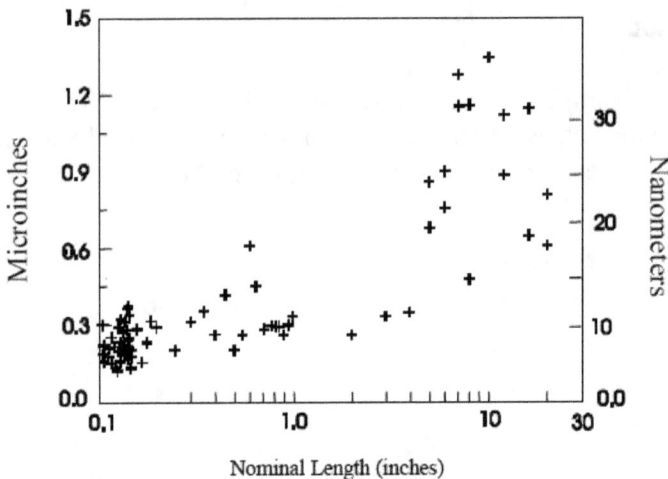

Figure 6.10. The variability (σ) of NIST master gauge blocks as a function of length.

The major components of the interferometric measurement uncertainty are shown in table 6.2. The errors can be divided into two classes: those which are correlated (the same for all measurements) and those which are uncorrelated (vary for each measurement). The classification depends on the measurement history. NIST gauge blocks which are measured over a long period of time, during which the thermometers and barometers are recalibrated. In this case the errors due to thermal expansion and refractive index corrections vary. For a customer block, measured 4 times over a 4 day period, they do not vary. Table 6.2 classifies the errors for both NIST masters and customer calibrations.

Table 6.2

Source of Uncertainty	Uncertainty Type for NIST Masters	Uncertainty Type for Customers
Reading error	uncorrelated	uncorrelated
Obliquity/Slit Correction	correlated	correlated
Wringing Film Thickness	uncorrelated	uncorrelated
Phase Correction	correlated	correlated
Thermal Expansion Correction		
Temperature Measurement	uncorrelated	correlated
Expansion Coefficient Value	correlated	correlated
Index of Refraction		
Edlén Equation	correlated	correlated
Environmental Measurements	uncorrelated	uncorrelated
Laser Frequency	uncorrelated	correlated

The interferometric history provides a statistical sample of all of the uncorrelated errors. Using the information from figure 6.10 we obtain an estimate of the uncorrelated errors to be:

$$U_u \, (\mu m) \sim 0.008 + 0.03 \times L \quad (L \text{ in meters}) \qquad (6.20)$$

Multiple measurements of a gauge block can reduce this figure; if n measurements are made the uncertainty of the average length becomes $\tilde{U_u}/n$. Unfortunately, the correlated errors are not sampled by repeated measurements and are not reduced by multiple measurements. An estimate of the correlated uncertainties for NIST master gauge block calibrations is:

$$U_c \, (\mu m) \sim 0.007 + 0.07 \times L \quad (L \text{ in meters}) \qquad (6.21)$$

Because customer blocks have more correlated errors and more importantly do not have measured thermal expansion coefficients, the use of multiple measurements is less effective in reducing measurement uncertainty. For most customers, interferometric measurements are not cost-effective since the uncertainty for a three wring interferometric measurement is not significantly lower than the mechanical comparison measurement, but costs twice as much. If, however, a customer sends in the same blocks for interferometric measurement over a number of years and uses the accumulated history then the uncertainty of the blocks will approach that of NIST master blocks.

6.8 Multiple wavelength interferometry

If we think of gauge block interferometry as the measurement of height with a ruler marked in units of $\lambda/2$, we quickly realize that our ruler has units unlike a normal ruler. The lines which show a distance of $\lambda/2$ are there but there are no numbers to tell us which mark we are looking at. This ruler can be used if we know the length of the block to better than $\lambda/4$ since we can use this distance and make the small correction (less than $\lambda/4$) found by interpolating between the lines on our scale. Suppose, however, we do not know the length this well ahead of time.

The correct method in this case is to use a number of rulers with different and incommensurate scales. In interferometry we can use several light sources ranging from blue to red to provide these different scales. A simple example is shown in figure 6.11.

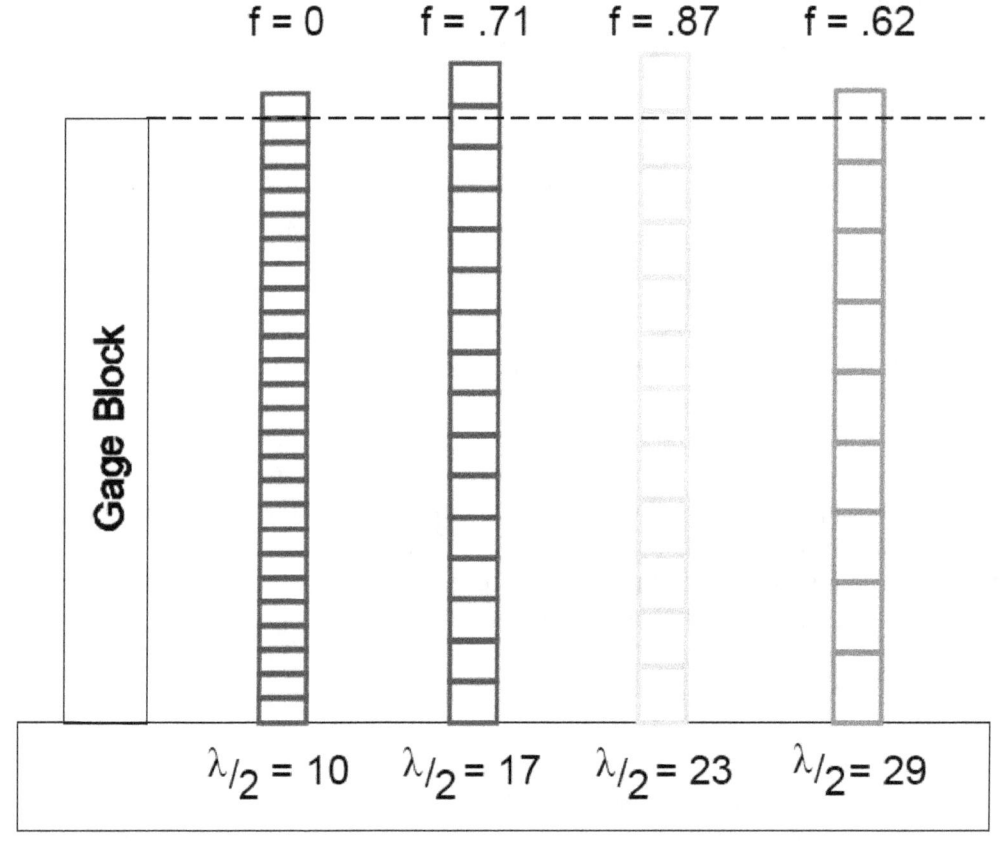

Figure 6.11. Four color interferometry is equivalent to measuring a length with four scales of known pitch which have indeterminate zeros.

For each scale (color) there will be a fringe fraction f_i. With only one color the block might have any length satisfying the formula:

$$L = (n_1 + f_1)*\lambda_1/2 \quad \text{where } n_1 \text{ is any integer.} \tag{6.22}$$

If we look at a second color, there will be another fringe fraction f_2. It also will be consistent with any block length satisfying the formula:

$$L = (n_2 + f_2)*\lambda_2/2 \quad \text{where } n_2 \text{ is any integer.} \tag{6.23}$$

However, we have gained something, since the number of lengths which satisfy **both** relations is very much reduced, and in fact are considerably further apart than $\lambda/2$. With more colors the number of possible matches are further reduced until a knowledge of the length of the block to the nearest centimeter or more is sufficient to determine the exact length.

In theory, of course, knowing the exact fringe fractions for two colors is sufficient to know any length since the two wavelengths are not commensurate. In practice, our knowledge of the fringe fractions is limited by the sensitivity and reproducibility of our equipment. In practice, 1/20 of a fringe is a conservative estimate for the fringe fraction uncertainty. A complete analysis of the effects of the uncertainty and choice of wavelengths on multicolor interferometry is given by Tilford [49].

Before the general availability of computers, the analysis of multicolor interferometry was a time consuming task [50]. There was a large effort made to produce calculation aids in the form of books of fringe fractions for each popular source wavelength, correction tables for the index of refraction, and even fringe fraction coincidence rules built somewhat like a slide rule. Since the advent of computers it is much easier to take the brute force approach since the calculations are quite simple for the computer.

Figure 6.11 shows a graphical output of the NIST multicolor interferometry program using a cadmium light source. The program calculates the actual wavelengths for each color using the environmental factors (air temperature, pressure and humidity). Then, using the observed fringe fraction, shows the possible lengths of the gauge block which are near the nominal length for each color. Note that the possible lengths are shown as small bars, with their width corresponding to the uncertainty in the fringe fraction.

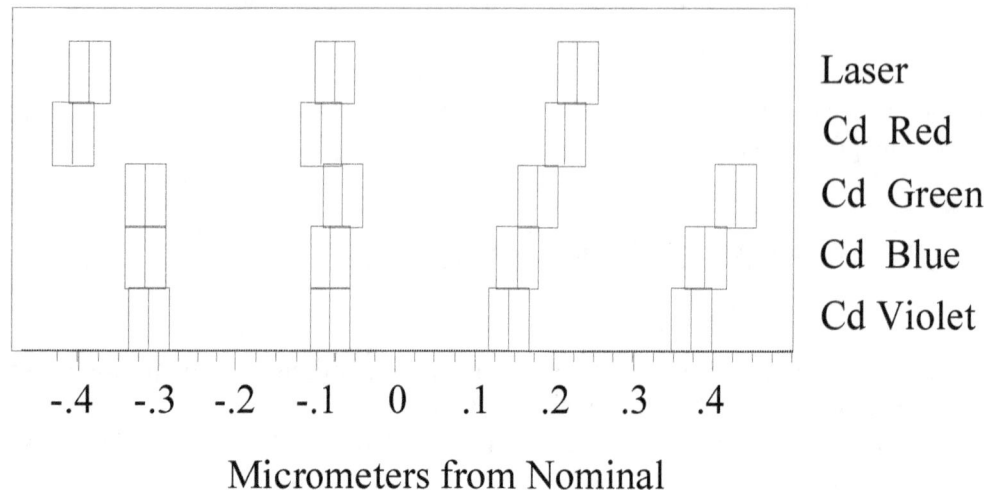

Figure 6.12. Graphical output of the NIST multicolor interferometry program. The length at which all the colors match within 0.030 μm is assumed to be the true gauge block length. The match is at -0.09 μm in this example.

The length where all of the fringes overlap is the actual length of the block. If all the fringes do not overlap in the set with the best fit, the inconsistency is taken as evidence of an operator error and the block is re-measured. The computer is also programmed to examine the data and decide if there is a length reasonably close to the nominal length for which the different wavelengths agree to a given tolerance. As a rule of thumb, all of the wavelengths should agree to better than 0.030 μm to be acceptable.

Analytic methods for analyzing multicolor interferometry have also been developed [49]. Our implementations of these types of methods have not performed well. The problem is probably that the span of wavelengths available, being restricted to the visible, is not wide enough and the fringe fraction measurement not precise enough for the algorithms to work unambiguously.

6.9 Use of the Linescale Interferometer for End Standard Calibration

There are a number of methods to calibrate a gauge block of completely unknown length. The multiple wavelength interferometry of the previous section is used extensively, but has the limitation that most atomic sources have very limited coherence lengths, usually under 25 mm. The method can be used by measuring a set of blocks against each other in a sequence to generate the longest length. For example, for a 10 inch block, a 1 inch block can be measured absolutely followed by differential measurements of a 2 inch block with the 1 inch block, a 3 inch block with the 2, a 4 inch block with the 3, a 5 inch block with the 4, and the 10 inch block with the 2, 3 and 5 inch blocks wrung together. Needless to say this method is tedious and involves the uncertainties of a large number of measurements.

Another, simpler, method is to convert a long end standard into a line standard and measure it with an instrument designed to measure or compare line scales (for example, meter bars) [51]. The NIST linescale interferometer, shown schematically below, is generally used to check our master blocks over 250 mm long to assure that the length is known within the 1/2 fringe needed for single wavelength interferometry.

The linescale interferometer consists of a 2 m long waybed which moves a scale up to a meter in length, under a microscope. An automated photoelectric microscope, sends a servo signal to the machine controller which moves the scale so that the graduation is at a null position on the microscope field of view. A laser interferometer measures the distances between the marks on the scale via a corner cube attached to one end of the scale support. This system is described in detail elsewhere [52,53].

To measure an end standard, two small gauge blocks that have linescale graduations on one side, are wrung to the ends of the end standard, as shown in figure 6.13. This "scale" is then measured on the linescale interferometer. The gauge blocks are then removed from the end standard and wrung together, forming a short scale. This "scale" is also measured on the interferometer. The difference in length between the two measurements is the physical distance between the end faces of the end standard plus one wringing film. This distance is the defined length of the end standard.

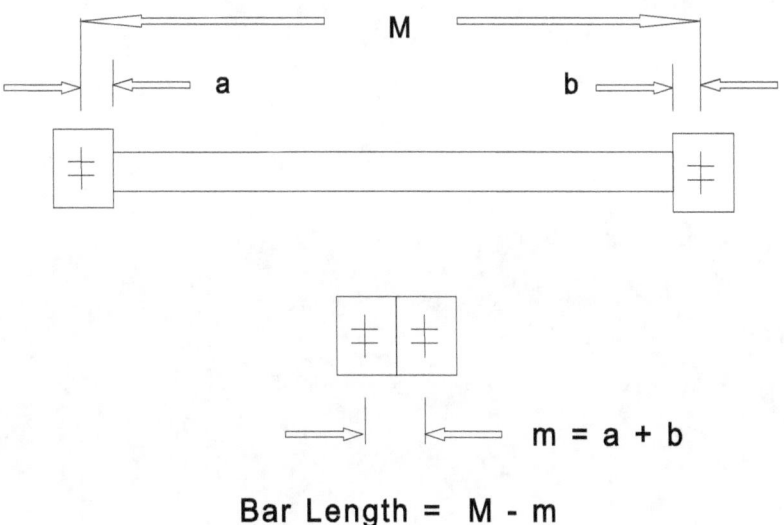

Figure 6.13. Two small gauge blocks with linescale graduations on one side are wrung to the ends of the end standard, allowing the end standard to be measured as a linescale.

The only significant problem with this method is that it is not a platen to gauge point measurement like a normal interferometric measurement. If the end standard faces are not flat and parallel the measurement will not give the exact same length, although knowledge of the parallelism and flatness will allow corrections to be made. Since the method is only used to determine the length within 1/2 fringe of the true length this correction is seldom needed.

7. References

[1] C.G. Peters and H.S. Boyd, "Interference methods for standardizing and testing precision gauge blocks," Scientific Papers of the Bureau of Standards, Vol. 17, p.691 (1922).

[2] Beers, J.S. "A Gauge Block Measurement Process Using Single Wavelength Interferometry," NBS Monograph 152, 1975.

[3] Tucker, C.D. "Preparations for Gauge Block Comparison Measurements," NBSIR 74-523.

[4] Beers, J.S. and C.D. Tucker. "Intercomparison Procedures for Gauge Blocks Using Electromechanical Comparators," NBSIR 76-979.

[5] Cameron, J.M. and G.E Hailes. "Designs for the Calibration of Small Groups of Standards in the Presence of Drift," NBS Technical Note 844, 1974.

[6] Klein, Herbert A., The Science of Measurement, Dover Publications, 1988.

[7] Galyer, J.F.W. and C.R. Shotbolt, Metrology for Engineers, Cassel & Company, Ltd., London, 1964.

[8] "Documents Concerning the New Definition of the Meter," Metrologia, Vol. 19, 1984.

[9] "Use of the International Inch for Reporting Lengths of Gauge Blocks," National Bureau of Standards (U.S.) Letter Circular LC-1033, May, 1959.

[10] T.K.W. Althin, C.E. Johansson, 1864-1943, Stockolm, 1948.

[11] Cochrane, Rexmond C., AMeasures for Progress," National Bureau of Standards (U.S.), 1966.

[12] Federal Specification: Gauge Blocks and Accessories (Inch and Metric), Federal Specification GGG-G-15C, March 20, 1975.

[13] Precision Gauge Blocks for Length Measurement (Through 20 in. and 500 mm), ANSI/ASME B89.1.9M-1984, The American Society of Mechanical Engineers, 1984.

[14] International Standard 3650, Gauge Blocks, First Edition, 1978-07-15, 1978..

[15] DIN 861, part 1, Gauge Blocks: Concepts, requirements, testing, January 1983.

[16] M.R. Meyerson, T.R. Young and W.R. Ney, "Gauge Blocks of Superior Stability: Initial Developments in Materials and Measurement," J. of Research of the National Bureau of Standards, Vol. 64C, No. 3, 1960.

[17] Meyerson, M.R., P.M. Giles and P.F. Newfeld, "Dimensional Stability of Gauge Block Materials," J. of Materials, Vol. 3, No. 4, 1968.

[18] Birch, K.P., "An automatic absolute interferometric dilatometer," J. Phys. E: Sci. Instrum, Vol. 20, 1987.

[19] J.W. Berthold, S.F. Jacobs and M.A. Norton, "Dimensional Stability of Fused Silica, Invar, and Several Ultra-low Thermal Expansion Materials," Metrologia, Vol. 13, pp. 9-16 (1977).

[20] C.W. Marshall and R.E. Maringer, Dimensional Instability, An Introduction, Pergamon Press, New York (1977).

[21] Hertz, H., "On the contact of elastic solids," English translation in Miscellaneous Papers, Macmillan, N.Y., 1896.

[22] Poole, S.P., "New method of measuring the diameters of balls to a high precision," Machinery, Vol 101, 1961.

[23] Norden, Nelson B., "On the Compression of a Cylinder in Contact with a Plane Surface," NBSIR 73-243, 1973.

[24] Puttock, M.J. and E.G. Thwaite, "Elastic Compression of Spheres and Cylinders at Point and Line Contact," National Standards Laboratory Technical Paper No. 25, CSIRO, 1969.

[25] Beers, John, and James E. Taylor, "Contact Deformation in Gauge Block Comparisons," NBS Technical Note 962, 1978

[26] Beyer-Helms, F., H. Darnedde, and G. Exner. "Langenstabilitat bei Raumtemperatur von Proben er Glaskeramik 'Zerodur'," Metrologia Vol. 21, p49-57 (1985).

[27] Berthold, J.W. III, S.F. Jacobs, and M.A. Norton. "Dimensional Stability of Fused Silica, Invar, and Several Ultra-low Thermal Expansion Materials," Metrologia, Vol. 13, p9-16 (1977).

[28] Justice, B., "Precision Measurements of the Dimensional Stability of Four Mirror Materials," Journal of Research of the National Bureau of Standards - A: Physics and Chemistry, Vol. 79A, No. 4, 1975.

[29] Bruce, C.F., Duffy, R.M., Applied Optics Vol.9, p743-747 (1970).

[30] Average of 1,2,3 and 4 inch steel master gauge blocks at N.I.S.T.

[31] Doiron, T., Stoup, J., Chaconas, G. and Snoots, P. "stability paper, SPIE".

[32] Eisenhart, Churchill, "Realistic Evaluation of the Precision and Accuracy of Instrument

Calibration Systems," Journal of Research of the National Bureau of Standards, Vol. 67C, No. 2, pp. 161-187, 1963.

[33] Croarkin, Carroll, "Measurement Assurance Programs, Part II: Development and Implementation," NBS Special Publication 676-II, 1984.

[34] ISO, "Guide to the Expression of Uncertainty in Measurement," October 1993.

[35] Taylor, Barry N. and Chris E. Kuyatt, "Guidelines for Evaluating and Expressing the Uncertainty of NIST Measurement Results," NIST Technical Note 1297, 1994 Edition, September 1994.

[36] Tan, A. and J.R. Miller, "Trend studies of standard and regular gauge block sets," Review of Scientific Instruments, V.62, No. 1, pp.233-237, 1991.

[37] C.F. Bruce, "The Effects of Collimation and Oblique Incidence in Length Interferometers. I," Australian Journal of Physics, Vol. 8, pp. 224-240 (1955).

[38] C.F. Bruce, "Obliquity Correction Curves for Use in Length Interferometry," J. of the Optical Society of America, Vol. 45, No. 12, pp. 1084-1085 (1955).

[39] B.S. Thornton, "The Effects of Collimation and Oblique Incidence in Length Interferometry," Australian Journal of Physics, Vol. 8, pp. 241-247 (1955).

[40] L. Miller, <u>Engineering Dimensional Metrology</u>, Edward Arnold, Ltd., London (1962).

[41] Schweitzer, W.G., et.al., "Description, Performance and Wavelengths of Iodine Stabilized Lasers," Applied Optics, Vol 12, 1973.

[42] Chartier, J.M., et. al., "Intercomparison of Northern European $^{127}I_2$ - Stabilized He-Ne Lasers at λ = 633 nm," Metrologia, Vol 29, 1992.

[43] Balhorn, R., H. Kunzmann and F. Lebowsky, AFrequency Stabilization of Internal-Mirror Helium-Neon Lasers,@ Applied Optics, Vol 11/4, April 1972.

[44] Mangum, B.W. and G.T. Furukawa, "Guidelines for Realizing the International Temperature Scale of 1990 (ITS-90)," NIST Technical Note 1265, National Institute of Standards and Tecnology, 1990.

[45] Edlen, B., "The Refractive Index of Air," Metrologia, Vol. 2, No. 2, 1966.

[46] Schellekens, P., G. WIlkening, F. Reinboth, M.J. Downs, K.P. Birch, and J. Spronck, "Measurements of the Refractive Index of Air Using Interference Refractometers," Metrologia, Vol. 22, 1986.

[47] Birch, K.P. and Downs, M.J., "The results of a comparison between calculated and measured values of the refractive index of air," J. Phys. E: Sci. Instrum., Vol. 21, pp. 694-695, 1988.

[48] Birch, K.P. and M.J. Downs, "Correction to the Updated Edlén Equation for the Refractive Index of Air," Metrologia, Vol. 31, 1994.

[49] C.R. Tilford, "Analytical Procedure for determining lengths from fractional fringes," Applied Optics, Vol. 16, No. 7, pp. 1857-1860 (1977).

[50] F.H. Rolt, <u>Gauges and Fine Measurements</u>, Macmillan and Co., Limited, 1929.

[51] Beers, J.S. and Kang B. Lee, "Interferometric measurement of length scales at the National Bureau of Standards," Precision Engineering, Vol. 4, No. 4, 1982.

[52] Beers, J.S., "Length Scale Measurement Procedures at the National Bureau of Standards," NBSIR 87-3625, 1987.

[53] Beers, John S. and William B. Penzes, "NIST Length Scale Interferometer Measurement Assurance," NISTIR 4998, 1992.

APPENDIX A. Drift Eliminating Designs for Non-simultaneous Comparison Calibrations

Introduction

The sources of variation in measurements are numerous. Some of the sources are truly random noise, 1/f noise in electronic circuits for example. Usually the "noise" of a measurement is actually due to uncontrolled systematic effects such as instability of the mechanical setup or variations in the conditions or procedures of the test. Many of these variations are random in the sense that they are describable by a normal distribution. Like true noise in the measurement system, the effects can be reduced by making additional measurements.

Another source of serious problems, which is not random, is drift in the instrument readings. This effect cannot be minimized by additional measurement because it is not generally pseudo-random, but a nearly monotonic shift in the readings. In dimensional metrology the most import cause of drift is thermal changes in the equipment during the test. In this paper we will demonstrate techniques to address this problem of instrument drift.

A simple example of the techniques for eliminating the effects of drift by looking at two different ways of comparing 2 gauge blocks, one standard (A) and one unknown (B).

$$\text{Scheme 1:} \quad A \ B \ A \ B$$

$$\text{Scheme 2:} \quad A \ B \ B \ A$$

Now let us suppose we make the measurements regularly spaced in time, 1 time unit apart, and there is an instrumental drift of Δ. The actual readings (y_i) from scheme 1 are:

$$y_1 = A \qquad \text{(A.1a)}$$
$$y_2 = B + \Delta \qquad \text{(A.1b)}$$
$$y_3 = A + 2\Delta \qquad \text{(A.1c)}$$
$$y_4 = B + 3\Delta \qquad \text{(A.1d)}$$

Solving for B in terms of A we get:

$$B = A - \frac{1}{2}(Y1 + Y3 - Y2 - Y4) - \Delta \qquad \text{(A.2)}$$

which depends on the drift rate Δ.

Now look at scheme 2. Under the identical conditions the readings are:

$$y_1 = A \quad \text{(A.3a)}$$
$$y_2 = B + \Delta \quad \text{(A.3b)}$$
$$y_3 = B + 2\Delta \quad \text{(A.3c)}$$
$$y_4 = A + 3\Delta \quad \text{(A.3d)}$$

Here we see that if we add the second and third readings and subtract the first and fourth readings we find that the Δ drops out:

$$B = A - \frac{1}{2}(Y1 + Y4 - Y2 - Y3) \quad \text{(A.4)}$$

Thus if the drift rate is constant - a fair approximation for most measurements if the time is properly restricted - the analysis both eliminates the drift and supplies a numerical approximation of the drift rate.

The calibration of a small number of "unknown" objects relative to one or two reference standards involves determining differences among the group of objects. Instrumental drift, due most often to temperature effects, can bias both the values assigned to the objects and the estimate of the effect of random errors. This appendix presents schedules for sequential measurements of differences that eliminate the bias from these sources and at the same time gives estimates of the magnitude of these extraneous components.

Previous works have [A1,A2] discussed schemes which eliminate the effects of drift for simultaneous comparisons of objects. For these types of measurements the difference between two objects is determined at one instant of time. Examples of these types of measurements are comparisons of masses with a double pan balance, comparison of standard voltage cells, and thermometers which are all placed in the same thermalizing environment. Many comparisons, especially those in dimensional metrology, cannot be done simultaneously. For example, using a gauge block comparator, the standard, control (check standard) and test blocks are moved one at a time under the measurement stylus. For these comparisons each measurement is made at a different time. Schemes which assume simultaneous measurements will, in fact, eliminate the drift from the analysis of the test objects but will produce a measurement variance which is drift dependent and an erroneous value for the drift, Δ.

In these calibration designs only differences between items are measured so that unless one or more of them are standards for which values are known, one cannot assign values for the remaining "unknown" items. Algebraically, one has a system of equations that is not of full rank and needs the value for one item or the sum of several items as the restraint to lead to a unique solution. The least squares method used in solving these equations has been presented [A3] and refined [A4] in the literature and will not be repeated in detail here. The analyses presented of particular measurement designs presented later in this paper conform to the method and notation presented in detail by Hughes [A3].

The schemes used as examples in this paper are those currently used at NIST for gauge block comparisons. In our calibrations a control (check standard) is always used to generate data for our measurement assurance plan [A5]. It is not necessary, however, to use a control in every measurement and the schemes presented can be used with any of the objects as the standard and the rest as unknowns. A number of schemes of various numbers of unknowns and measurements is presented in the appendix.

Calibration Designs

The term calibration design has been applied to experiments where only differences between nominally equal objects or groups of objects can be measured. Perhaps the simplest such experiment consists in measuring the differences between the two objects of the n(n-1) distinct pairings that can be formed from n objects. If the order is unimportant, X compared to Y is the negative of Y compared to X, there are only n(n-1)/2 distinct pairings. Of course only 1 measurement per unknown is needed to determine the unknown, but many more measurements are generally taken for statistical reasons. Ordinarily the order in which these measurements are made is of no consequence. However, when the response of the comparator is time dependent, attention to the order is important if one wished to minimize the effect of these changes.

When this effect can be adequately represented by a linear drift, it is possible to balance out the effect by proper ordering of the observations. The drift can be represented by the series, ... -3 , -2 , -1 , 0, 1, 2, 3 , ... if there are an odd number of comparisons and by ... -5/2 , -3/2 , -1/2 , 1/2 , 3/2 , 5/2 , ... if there are an even number of comparisons.

As and example let us take n=3. If we make all possible n(n-1)=6 comparisons we get a scheme like that below, denoting the three objects by A, B, C.

Observation	Measurement
m_1	A - 11/2 Δ
m_2	B - 9/2 Δ
m_3	C - 7/2 Δ
m_4	A - 5/2 Δ
m_5	B - 3/2 Δ
m_6	C - 1/2 Δ
m_7	A + 1/2 Δ
m_8	C + 3/2 Δ
m_9	B + 5/2 Δ
m_{10}	A + 7/2 Δ
m_{11}	C + 9/2 Δ
m_{12}	B + 11/2 Δ

(A.5)

If we analyze these measurements by pairs, in analogy to the weighing designs of Cameron we see that:

$$
\begin{array}{llll}
y_1 = m_1 - m_2 & = A - B - \Delta \\
y_2 = m_3 - m_4 & = C - A - \Delta \\
y_3 = m_5 - m_6 & = B - C - \Delta \\
y_4 = m_7 - m_8 & = A - C - \Delta \\
y_5 = m_9 - m_{10} & = B - A - \Delta \\
y_6 = m_{11} - m_{12} & = C - B - \Delta
\end{array}
\quad
\begin{array}{cccc}
A & B & C & \Delta \\
+1 & -1 &   & -1 \\
-1 &   & +1 & -1 \\
   & +1 & -1 & -1 \\
+1 &   & -1 & -1 \\
-1 & +1 &   & -1 \\
   & +1 & -1 & -1
\end{array}
\qquad (A.6)
$$

The notation used here, the plus and minus signs, indicate the items entering into the difference measurement. Thus, y_2 is a measurement of the difference between object C and object A.

Note the difference between the above table and that of simultaneous comparisons in reference 2 is that the drift column is constant. It is simple to see by inspection that the drift is balanced out since each object has two (+) and two (-) measurements and the drift column is constant. By extension, the effects of linear drift is eliminated in all complete block measurement schemes (those for which all objects are measured in all possible n(n-1) combinations).

Although all schemes in which each object has equal numbers of (+) and (-) measurements is drift eliminating, there are practical criteria which must be met for the scheme to work. First, the actual drift must be linear. For dimensional measurements the instrument drift is usually due to changes in temperature. The usefulness of drift eliminating designs depends on the stability of the thermal environment and the accuracy required in the calibration. In the NIST gauge block lab the environment is stable enough that the drift is linear at the 3 nm (0.1 µin) level over periods of 5 to 10 minutes. Our comparison plans are chosen so that the measurements can be made in this period. Secondly, each measurement must be made in about the same amount of time so that the measurements are made at fairly regular intervals. In a completely automated system this is simple, but with human operators there is a natural tendency to make measurements simpler and quicker is the opportunity presents itself. For example, if the scheme has a single block measured two or more times in succession it is tempting to measure the object without removing it from the apparatus, placing it in its normal resting position, and returning it to the apparatus for the next measurement.

Finally, the measurements of each block are spread as evenly as possible across the design. Suppose in the scheme above where each block is measured 4 times block, A is measured as the first measurement of y_1, y_2, y_3, and y_4. There is a tendency to leave block A near the measuring point rather than its normal resting position because it is used so often in the first part of the scheme. This allows block A to have a different thermal handling than the other blocks which can result in a thermal drift which is not the same as the other blocks.

Restraints

Since all of the measurements made in a calibration are relative comparisons, at least one value must be known to solve the system of equations. In the design of the last section, for example, if one has a single standard and two unknowns, the standard can be assigned to any one of the letters. (The same would be true of three standards and one unknown.) If there are two standards and one unknown, the choice of which pair of letters to assign for the standards is important in terms of minimizing the uncertainty in the unknown.

For full block designs (all possible comparisons are made) there is no difference which label is used for the standards or unknowns. For incomplete block designs the uncertainty of the results can depend on which letter the standard and unknowns are assigned. In these cases the customer blocks are assigned to minimize their variance and allow the larger variance for the measurement of the extra master (control).

This asymmetry occurs because every possible comparison between the four items has not been measured. For 4 objects there are 12 possible intercomparisons. If an 8 measurement scheme is used all three unknowns cannot be compared directly to the standard the same number of times. For example, two unknowns can be compared directly with the standard twice, but the other unknown will have no direct comparisons. This indirect comparison to the standard results in a slightly larger variance for the block compared indirectly. Complete block plans, which compare each block to every other block equal number of times, have no such asymmetry, and thus remove any restriction on the measurement position of the control.

Example: 4 block, 12 comparison, Single Restraint Design for NIST Gauge Block Calibration

The gauge block comparison scheme put into operation in 1989 consists of two standards blocks, denoted S and C, and two customer blocks to be calibrated, denoted X and Y. In order to decrease the random error of the comparison process a new scheme was devised consisting of all 12 possible comparisons between the four blocks. Because of continuing problems making penetration corrections, the scheme was designed to use either the S or C block as the restraint and the difference (S-C) as the control parameter. The S blocks are all steel, and are used as the restraint for all steel customer blocks. The C blocks are chrome carbide, and are used as the restraint for chrome and tungsten carbide blocks. The difference (S-C) is independent of the choice of restraint.

We chose a complete block scheme that assures that the design is drift eliminating, and the blocks can be assigned to the letters of the design arbitrarily. We chose (S-C) as the first comparison. Since there are a large number of ways to arrange the 12 measurements for a complete block design, we added two restrictions as a guide to choose a "best" design.

1. Since the blocks are measured one at a time, it was decided to avoid schemes which measured the same block two or more times consecutively. In the scheme presented earlier blocks D, A and C are all measured twice consecutively. There is a great temptation to not remove and replace the blocks under these conditions, and the scheme assumes that each measurement is made with the same motion and are evenly spaced in time. This repetition threatens both these assumptions.

2. We decided that schemes in which the six measurements of each block were spread out as

evenly as possible in time would be less likely to be affected by small non-linearities in the drift. For example, some schemes had one block finished its 6 measurements by the 8th comparison, leaving the final 1/3 of the comparisons with no sampling of that block.

The new scheme is as follows:

$$Y = \begin{vmatrix} Y_1 \\ Y_2 \\ Y_3 \\ Y_4 \\ Y_5 \\ Y_6 \\ Y_7 \\ Y_8 \\ Y_9 \\ Y_{10} \\ Y_{11} \\ Y_{12} \end{vmatrix} = \begin{vmatrix} S-C \\ Y-S \\ X-Y \\ C-S \\ C-X \\ Y-X \\ S-X \\ C-Y \\ S-Y \\ X-C \\ X-S \\ Y-X \end{vmatrix} \quad X = \begin{vmatrix} S & C & X & Y & \Delta \\ 1 & -1 & 0 & 0 & -1 \\ -1 & 0 & 0 & 1 & -1 \\ 0 & 0 & 1 & -1 & -1 \\ -1 & 1 & 0 & 0 & -1 \\ 0 & 1 & -1 & 0 & -1 \\ 0 & -1 & 0 & 1 & -1 \\ 1 & 0 & -1 & 0 & -1 \\ 0 & 1 & 0 & -1 & -1 \\ 1 & 0 & 0 & -1 & -1 \\ 0 & -1 & 1 & 0 & -1 \\ -1 & 0 & 1 & 0 & -1 \\ 0 & 0 & -1 & 1 & -1 \end{vmatrix} \quad (A.7)$$

When the S block is the restraint (S – L) the matrix equation to solve is:

$$A = \begin{vmatrix} X'X & a \\ a^t & 0 \end{vmatrix}^{-1} \cdot \begin{vmatrix} X'Y \\ L \end{vmatrix} \quad (A.8)$$

$$|a^t| = [\,1\ 0\ 0\ 0\,] \quad (A.9)$$

$$|C_{ij}| = \begin{vmatrix} X'X & a \\ a^t & 0 \end{vmatrix} = \begin{vmatrix} 6 & -2 & -2 & -2 & 0 & 1 \\ -2 & 6 & -2 & -2 & 0 & 0 \\ -2 & -2 & 6 & -2 & 0 & 0 \\ -2 & -2 & -2 & 6 & 0 & 0 \\ 0 & 0 & 0 & 0 & 12 & 0 \\ 1 & 0 & 0 & 0 & 0 & 0 \end{vmatrix} \quad (A.10)$$

$$|C_{ij}|^{-1} = \left(\frac{1}{24}\right)\begin{vmatrix} 0 & 0 & 0 & 0 & 0 & 24 \\ 0 & 6 & 3 & 3 & 0 & 24 \\ 0 & 3 & 6 & 3 & 0 & 24 \\ 0 & 3 & 3 & 6 & 0 & 24 \\ 0 & 0 & 0 & 0 & 2 & 0 \\ 24 & 24 & 24 & 24 & 0 & 0 \end{vmatrix} \quad \text{(A.11)}$$

Using S as the restraint, the solution to the equations is:

$$S = L \tag{A.12a}$$

$$C = (1/8)(-2y_1 + y_2 + 2y_4 + y_5 - y_6 - y_7 + y_8 - y_9 - y_{10} + y_{11}) + L \tag{A.12b}$$

$$X = (1/8)(-y_1 + y_2 + y_3 + y_4 - y_5 - 2y_7 - y_9 + y_{10} + 2y_{11} - y_{12}) + L \tag{A.12c}$$

$$Y = (1/8)(-y_1 + 2y_2 - y_3 + y_4 + y_6 - y_7 - y_8 - 2y_9 + y_{11} + y_{12}) + L \tag{A.12d}$$

$$\Delta = (-1/12)(y_1 + y_2 + y_3 + y_4 + y_5 + y_6 + y_7 + y_8 + y_9 + y_{10} + y_{11} + y_{12}) \tag{A.12e}$$

The deviations, $d_1, d_2, ..., d_{12}$ can be determined from the equations above, or can be calculated directly using matrix methods. For example,

These deviations provide the information needed to obtain a value, s, which is the experiment's value for the short term process standard deviation, or within standard deviation σ_w.

$$s^2 = \sum_{i=0}^{n}((Y_i - \sum_{r=0}^{m}(A_r \times X_{tr}))^2)/(n-m+1) \tag{A.13}$$

The number of degrees of freedom results from taking the number of observations (n=12) less the number of unknowns (m=5; S, C, X, Y, Δ), and then adding one for the restraint. Because of the complete block structure (all 12 possible combinations measured) all of the standard deviations are the same:

$$\sigma_w \approx \sqrt{B_{ii}} s = (1/2)s \tag{A.14}$$

Process Control: F - Test

Continued monitoring of the measurement process is required to assure that predictions based on the accepted values for process parameters are still valid. For gauge block calibration at NIST, the process is monitored for precision by comparison of the observed standard deviation, σ_w, to the average of previous values. For this purpose the value of σ_w is recorded for every calibration done, and is periodically analyzed to provide an updated value of the accepted process σ_w for each gauge block size.

The comparison is made using the F distribution, which governs the comparison of variances. The ratio of the variances s^2 (derived from the model fit to each calibration) and σ_w^2 derived from the history is compared to the critical value $F(8,\infty,\alpha)$, which is the α probability point of the F distribution for degrees of freedom 8 and ∞. For calibrations at NIST, α is chosen as 0.01 to give $F(8,\infty,.01) = 2.5$.

$$F = \frac{s^2_{obs}}{\sigma_t^2} < 2.5 \tag{A.15}$$

If this condition is violated the calibration fails, and is repeated. If the calibration fails more than once the test blocks are re-inspected and the instrument checked and recalibrated. All calibrations, pass or fail, are entered into the history file.

Process Control: T - Test

At NIST a control measurement is made with each calibration by using two known master blocks in each calibration. One of the master blocks is steel and the other chrome carbide. When a customer block is steel the steel master is used as the restraint, and when a customer block is carbide, the carbide master is used as the restraint. The use of a control measurement for calibrations is necessary in order to provide assurance of the continuing accuracy of the measurements. The F-test, while providing some process control, only attempts to control the repeatability of the process, not the accuracy. The use of a control is also the easiest method to find the long term variability of the measurement process.

While the use of a control in each calibration is not absolutely necessary, the practice is highly recommended. There are systems that use intermittent tests, for example measurements of a control set once a week. This is a good strategy for automated systems because the chance of block to block operator errors is small. For manual measurements the process variability, and of course the occurrence of operator error is much higher.

The check for systematic error is given by comparison of the observed value of the difference between the standard and control blocks. If S is the standard it becomes the restraint, and if A is used as the control (S-A) is the control parameter for the calibration. This observed control is recorded for every calibration, and is used to periodically used to update the accepted, or average value, of the control. The process control step involves the comparison of the observed value of the control to the accepted (historical) value. The comparison is made using the Student t-distribution.

The control test demands that the observed difference between the control and its accepted value be less than 3 times the accepted long term standard deviation, σ_t, of the calibration process. This value of the t-distribution implies that a good calibration will not be rejected with a confidence level of 99.7%.

$$T = \frac{(A_1 - A_2)_{obs} - (A_1 - A_2)_{acc}}{\sigma_t} < 3 \qquad (A.16)$$

The value of σ_t is obtained directly from the sequence of values of (S-A) arising in regular calibrations. The recorded (S-C) values are fit to a straight line, and the square root of the variance of the deviations from this line is used as the total standard deviation, (σ_t).

If both the precision (F-test) and accuracy (t-test) criteria are satisfied, the process is regarded as being "in control" and values for the unknown, X, and its associated uncertainty are regarded as valid. Failure on either criterion is an "out-of-control" signal and the measurements are repeated.

The value for drift serves as an indicator of possible trouble if it changes markedly from its usual range of values. However, because any linear drift is balanced out, a change in the value does not of itself invalidate the result.

Conclusion

The choice of the order of comparisons is an important facet of calibrations, in particular if chosen properly the comparison scheme can be made immune to linear drifts in the measurement equipment. The idea of making a measurement scheme robust is a powerful one. What is needed to implement the idea is an understanding of the sources of variability in the measurement system. While such a study is sometimes difficult and time consuming because of the lack of reference material about many fields of metrology, the NIST experience has been that such efforts are rewarded with measurement procedures which, for about the same amount of effort, produce higher accuracy.

References

[A1] J.M. Cameron, M.C. Croarkin, and R.C. Raybold, "Designs for the Calibration of Standards of Mass," NBS Technical Note 952, 1977.

[A2] Cameron, J.M. and G.E Hailes. "Designs for the Calibration of Small Groups of Standards in the Presence of Drift," NBS Technical Note 844, 1974.

[A3] C.G. Hughes, III and H.A. Musk. "A Least Squares Method for Analysis of Pair Comparison Measurements," Metrologia, Volume 8, pp. 109-113 (1972).

[A4] C. Croarkin, "An Extended Error Model for Comparison Calibration," Metrologia, Volume 26, pp. 107-113, 1989.

[A5] C. Croarkin, "Measurement Assurance Programs, Part II: Development and Implementation," NBS Special Publication 676-II, 1984.

A Selection of Other Drift Eliminating Designs

The following designs can be used with or without a control block. The standard block is denoted S, and the unknown blocks A, B, C, etc. If a check standard block is used it can be assigned to any of the unknown block positions. The name of the design is simply the number of blocks in the design and the total number of comparisons made.

3-6 Design

(One master block,
2 unknowns
4 measurements each)

$y_1 = S - A$
$y_2 = B - S$
$y_3 = A - B$
$y_4 = A - S$
$y_5 = B - A$
$y_6 = S - B$

3-9 Design

(One master block,
2 unknowns,
6 measurements each)

$y_1 = S - A$
$y_2 = B - A$
$y_3 = S - B$
$y_4 = A - S$
$y_5 = B - S$
$y_6 = A - B$
$y_7 = A - S$
$y_8 = B - A$
$y_9 = S - B$

4-8 Design

(One master block,
3 unknowns,
4 measurements each)

$y_1 = S - A$
$y_2 = B - C$
$y_3 = C - S$
$y_4 = A - B$
$y_5 = A - S$
$y_6 = C - B$
$y_7 = S - C$
$y_8 = B - A$

4-12 Design

(One master block,
3 unknowns
6 measurements each)

$y_1 = S - A$
$y_2 = C - S$
$y_3 = B - C$
$y_4 = A - S$
$y_5 = A - B$
$y_6 = C - A$
$y_7 = S - B$
$y_8 = A - C$
$y_9 = S - C$
$y_{10} = B - A$
$y_{11} = B - S$
$y_{12} = C - B$

5-10 Design

(One master block,
4 unknowns,
4 measurements each)

$y_1 = S - A$
$y_2 = D - C$
$y_3 = S - B$
$y_4 = D - A$
$y_5 = C - B$
$y_6 = A - C$
$y_7 = B - S$
$y_8 = B - D$
$y_9 = C - S$
$y_{10} = A - D$

6-12 Design

(One master block,
5 unknowns,
4 measurements each)

$y_1 = S - A$
$y_2 = D - C$
$y_3 = E - B$
$y_4 = E - D$
$y_5 = C - A$
$y_6 = B - C$
$y_7 = S - E$
$y_8 = A - D$
$y_9 = A - B$
$y_{10} = D - S$
$y_{11} = B - E$
$y_{12} = C - S$

7-14 Design
(One master block,
6 unknowns
4 measurements each)

$y_1 = S - A$
$y_2 = E - C$
$y_3 = B - D$
$y_4 = A - F$
$y_5 = S - E$
$y_6 = D - B$
$y_7 = A - C$
$y_8 = B - F$
$y_9 = D - E$
$y_{10} = F - S$
$y_{11} = E - A$
$y_{12} = C - B$
$y_{13} = C - S$
$y_{14} = F - D$

8-16 Design
(One master block,
7 unknowns,
4 measurements each)

$y_1 = S - A$
$y_2 = E - G$
$y_3 = F - C$
$y_4 = D - S$
$y_5 = B - E$
$y_6 = G - F$
$y_7 = C - B$
$y_8 = E - A$
$y_9 = F - D$
$y_{10} = C - S$
$y_{11} = A - G$
$y_{12} = D - B$
$y_{13} = C - S$
$y_{14} = G - C$
$y_{15} = B - D$
$y_{16} = A - F$

9-18 Design
(One master block,
7 unknowns,
4 measurements each)

$y_1 = S - A$
$y_2 = H - F$
$y_3 = A - B$
$y_4 = D - C$
$y_5 = E - G$
$y_6 = C - A$
$y_7 = B - F$
$y_8 = G - H$
$y_9 = D - S$
$y_{10} = C - E$
$Y_{11} = H - S$
$y_{12} = G - D$
$y_{13} = C - S$
$y_{14} = A - C$
$y_{15} = F - D$
$y_{16} = S - H$
$y_{17} = E - B$
$y_{18} = F - G$

10-20 Design

(One master block,
9 unknowns
4 measurements each)

$y_1 = S - A$
$y_2 = F - G$
$y_3 = I - C$
$y_4 = D - E$
$y_5 = A - H$
$y_6 = B - C$
$y_7 = G - H$
$y_8 = I - S$
$y_9 = E - F$
$y_{10} = H - I$
$y_{11} = D - F$
$y_{12} = A - B$
$y_{13} = C - I$
$y_{14} = H - E$
$y_{15} = B - G$
$y_{16} = S - D$
$y_{17} = F - B$
$y_{18} = C - D$
$y_{19} = G - S$
$y_{20} = E - A$

11-22 Design

(One master block,
10 unknowns,
4 measurements each)

$y_1 = S - A$
$y_2 = D - E$
$y_3 = G - I$
$y_4 = C - H$
$y_5 = A - B$
$y_6 = I - J$
$y_7 = H - F$
$y_8 = D - S$
$y_9 = B - C$
$y_{10} = S - E$
$y_{11} = A - G$
$y_{12} = F - B$
$y_{13} = E - F$
$y_{14} = J - A$
$y_{15} = C - D$
$y_{16} = H - J$
$y_{17} = F - G$
$y_{18} = I - S$
$y_{19} = B - H$
$y_{20} = G - D$
$y_{21} = J - C$
$y_{22} = E - I$

Appendix B: Wringing Films

> In recent years it has been found possible to polish plane surfaces of hardened steel to a degree of accuracy which had previously been approached only in the finest optical work, and to produce steel blocks in the form of end gauges which can be made to adhere or "wring" together in combinations. Considerable interest has been aroused by the fact that these blocks will often cling together with such tenacity that a far greater force must be employed to separate them than would be required if the adhesion were solely due to atmospheric pressure. It is proposed in this paper to examine the various causes which produce this adhesion: firstly, showing that by far the greater portion of the effect is due to the presence of a liquid film between the faces of the steel; and , secondly, endeavoring to account for the force which can be resisted by such a film.

Thus began the article "The Adherence of Flat Surfaces" by H.M. Budgett in 1912 [B1], the first scientific attack on the problem of gauge block wringing films. Unfortunately for those wishing tidy solutions, the field has not progressed much since 1912. The work since then has, of course, added much to our qualitative understanding of various phenomena associated with wringing, but there is still no clear quantitative or predictive model of wringing film thickness or its stability in time. In this appendix we will only describe some properties of wringing films, and make recommendations about strategies to minimize problems due to film variations.

Physics of Wringing Films

What causes wrung gauge blocks to stick together? The earliest conjectures were that sliding blocks together squeezed the air out, creating a vacuum. This view was shown to be wrong as early as 1912 by Budgett [B1] but still manages to creep into even modern textbooks [B2]. It is probable that wringing is due to a number of forces, the relative strengths of which depend on the exact nature of the block surface and the liquid adhering to the surface. The known facts about wringing are summarized below.

1. The force of adhesion between blocks can be up to 300 N (75 lb). The force of the atmosphere, 101 KPa (14 psi), is much weaker than an average wring, and studies have shown that there is no significant vacuum between the blocks.

2. There is some metal-metal contact between the blocks, although too small for a significant metallic bond to form. Wrung gauge blocks show an electrical resistance of about 0.003Ω [B3] that corresponds to an area of contact of 10^{-5} cm^2.

3. The average wringing film thickness depends on the fluid and surface finishes, as well as the amount of time blocks are left wrung together. Generally the thickness is about 10 nm (0.4 µin), but some wrings will be over 25 nm (1 µin) and some less than 0. (Yes, less than zero.) [B3,B4.B5,B6]

4. The fluid between blocks seems to provide much of the cohesive force. No matter how a block is cleaned, there will be some small amount of adsorbed water vapor. The normal

wringing procedure, of course, adds minute amounts of grease which allows a more consistent wringing force. The force exerted by the fluid is of two types. Fluid, trapped in the very small space between blocks, has internal bonds that resist being pulled apart. The fluid also has a surface tension that tends to pull blocks together. Both of these forces are large enough to provide the observed adhesion of gauge blocks.

5. The thickness of the wringing film is not stable, but evolves over time. First changes are due to thermal relaxation, since some heat is transferred from the technician's hands during wringing. Later, after the blocks have come to thermal equilibrium, the wring will still change slowly. Over a period of days a wring can grow, shrink or even complicated combinations of growth and shrinkage [B5,B6].

6. As a new block is wrung repeatedly the film thickness tends to shrink. This is due to mechanical wear of the high points of the gauge block surface [B5,B6].

7. As blocks become worn and scratched the wringing process becomes more erratic, until they do not wring well. At this point the blocks should be retired.

There may never be a definitive physical description for gauge block wringing. Besides the papers mentioned above, which span 60 years, there was a large project at the National Bureau of Standards during the 1960's. This program studied wringing films by a number of means, including ellipsometry [B8]. The results were very much in line with the 7 points given above, i.e., on a practical level we can describe the length properties of wringing films but lack a deeper understanding of the physics involved in the process.

Fortunately, standards writers have understood this problem and included the length of one wringing film in the defined block length. This relieves us of determining the film thickness separately since it is automatically included whenever the block is measured interferometrically. There is some uncertainty left for blocks that are measured by mechanical comparison, since the length of the master block wringing film is assumed to be the same as the unknown block. This uncertainty is probably less than 5 nm (.2 μin) for blocks in good condition.

REFERENCES

[B1]   "H.M. Budgett, "The Adherence of Flat Surfaces," Proceedings of the Royal Society, Vol. 86A, pp. 25-36 (1912).

[B2]   D.M. Anthony, Engineering Metrology, Peragamon Press, New York, N.Y. (1986).

[B3]   C.F. Bruce and B.S. Thornton, "Adhesion and Contact Error in Length Metrology," Journal of Applied Physics, Vol. 17, No.8 pp. 853-858 (1956).

[B4]   C.G. Peters and H.S. Boyd, "Interference methods for standardizing and testing precision gauge blocks," Scientific Papers of the Bureau of Standards, Vol. 17, p.691 (1922).

[B5] F.H. Rolt and H. Barrell, "Contact of Flat Surfaces," Proc. of the Royal Society (London), Vol. 106A, pp. 401-425 (1927).

[B6] G.J. Siddall and P.C. Willey, "Flat-surface wringing and contact error variability," J. of Physics D: Applied Physics, Vol. 3, pp. 8-28 (1970).

[B7] J.M. Fath, "Determination of the Total Phase, Basic Plus Surface Finish Effect, for Gauge Blocks," National Bureau of Standards Report 9819 (1967).

Appendix C. Phase Shifts in Gauge block Interferometry

When light reflects from a surface there is a shift in phase; the reflected light appears to have reflected from a plane below the mechanical surface, as shown in figure C. Many years ago, when surface finishes were rougher, the main component of phase differences between metal surfaces was due to surface finish. Current manufacturing practice has reduced this component to a small part of the phase shift. There are still significant phase shift differences between blocks of different manufacturers and different materials.

Nonconducting materials, such as quartz and glass have phase shifts of 180°. Electrical conductors will have phase shifts less than 180° by 10° to 30°. The theoretical foundation for this shift is given in Ditchburn [C1]. From an understanding of electromagnetic behavior of electrons near the metal surface the phase shift could be calculated. There have been such calculations using the Drude model for electrons. Unfortunately this classical model of electrons in metals does not describe electron behavior at very high frequencies, and is only useful for calculations of phase shifts for wavelengths in the infrared and beyond. There have been no successful calculations for phase shifts at higher (visible light) frequencies, and there is no currently available theory to predict phase shifts from any other measurable attributes of a metal.

Given this fact, phase shifts must be measured indirectly. There is one traditional method to measure phase shift, the "slave block" method [C2]. In this method an auxiliary block, called the slave block, is used to help find the phase shift difference between a block and a platen. The method consists of two steps, shown schematically in figure C2 and C3.

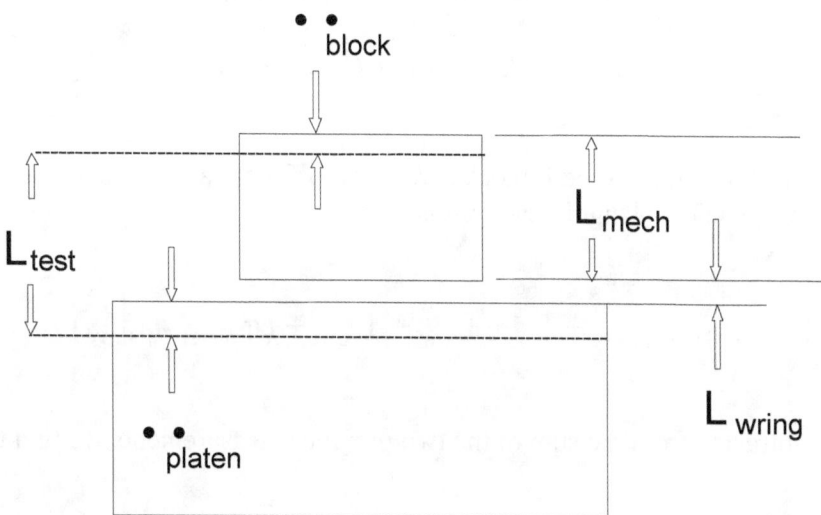

Figure C. The interferometric length, L_{test}, includes the mechanical length, the wringing film thickness and the phase change at each surface.

Step 1. The test and slave blocks are wrung down to the same platen and measured independently. The two lengths measured consist of the mechanical length of the block, the wringing film and the phase changes at the top of the block and platen, as in C.

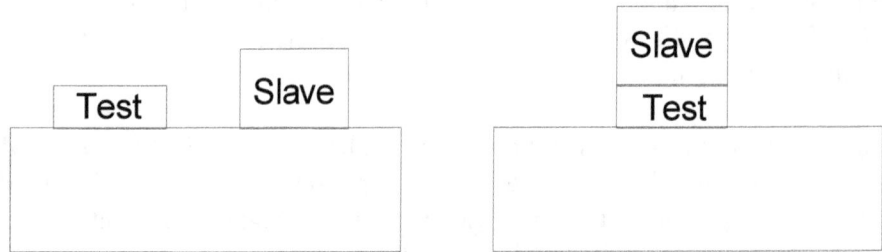

Figure C2. Measurements for determining the phase shift difference between a block and platen by the slave block method.

The general formula for the measured length of a wrung block is:

$$L_{test} = L_{mechanical} + L_{wring} + L_{platen\ phase} - L_{block\ phase} \tag{C.1}$$

For the test and slave blocks the formulas are:

$$L_{test} = L_t + L_{t,w} + (\varphi_{platen} - \varphi_{test}) \tag{C.2}$$

$$L_{slave} = L_s + L_{s,w} + (\varphi_{platen} - \varphi_{slave}) \tag{C.3}$$

Step 2. Either the slave block or both blocks are taken off the platen, cleaned, and rewrung as a stack on the platen. The new length measured is:

$$L_{test+slave} = L_t + L_s + L_{t,w} + L_{s,w} + (\varphi_{platen} - \varphi_{slave}) \tag{C.4}$$

If this result is subtracted from the sum of the two previous measurements we find that:

$$L_{test+slave} - L_{test} - L_{slave} = (\varphi_{test} - \varphi_{platen}) \tag{C.5}$$

The weakness of this method is the uncertainty of the measurements. The uncertainty of one measurement of a wrung gauge block is about 30 nm (1.2 µin). Since the phase measurement depends on the difference between 3 measurements, the phase measurement is uncertain to about 3 x uncertainty of one measurement, or about 50 nm. Since the phase difference between block and platen is generally about 20 nm, the uncertainty is larger than the effect. To reduce the uncertainty a large number of measurements must be made, generally between 50 and 100. This is, of course, very time consuming.

Table C shows a typical result of the slave block method for blocks wrung to quartz. The uncertainty is, unfortunately large; we estimate the standard uncertainty (95% confidence level) to be 8 nm.

Table C

Corrections to Nominal Lengths and Phase in Nanometers

Manufacturer	Material	Measured Phase (nanometers)
DoAll	steel	25
C.E. Johansson	steel	50
Matrix	steel	50
Pratt & Whitney	steel	50
Webber	steel	55
Webber	chrome carbide	30
Mitutoyo	steel	40

We have made attempts to measure the phase for the ceramic (zirconia) gauge blocks but have had problems with residual stresses in the block stack. We have measured the phase to be 35 nm, but the geometry leads us to suspect that the measured phase might be a distortion of the block rather than the interferometric phase.

There have been a number of attempts to measure phase by other means [C3], but none have been particularly successful. One method, using Newton's rings formed by a spherical quartz surface in contact with a metal surface is under investigation [C4]. With computer vision systems to analyze fringe patterns, this method may become useful in the future.

As a rule, blocks from a single manufacturer have the same phase since the material and lapping method (and thus surface finishes) are the same for all blocks. Blocks from the same manufacturer but with different materials, or the same material from different manufacturers, generally have different phases. One way to reduce phase problems is to restrict your master blocks to one manufacturer and one material.

Another way to reduce effects of phase change is to use blocks and platens of the same material and manufacturer. While this seems simple in principle, it is not as easy in practice. Most manufacturers of gauge blocks do not make gauge block platens, and even when they do, lapping procedures for large platens are not always the same as procedures for blocks. At NIST we currently do not have any block/platen combinations which have matched phase.

REFERENCES

[C1]   R.W. Ditchburn, Light, Interscience Publications, 2nd. edition, 1963.

[C2]   Miller, L., Engineering Dimensional Metrology, Edward Arnold, Ltd., London, 1962.

[C3]   J.M. Fath, "Determination of the Total Phase, Basic Plus Surface Finish Effect, for Gauge Blocks," National Bureau of Standards Report 9819 (1967).

[C4]   K.G. Birch and R. Li, "A Practical Determination of the Phase Change at Reflection," IMEKO Symposium on Laser Applications in Precision Measurement, pp. 5-25, Budapest, 1986.

Appendix D. Deformation Corrections

Whenever two materials are forced into contact there is an elastic deformation. For gauge block comparators the contact is between a spherical probe and the flat gauge block surface. The following formula, from a CSIRO Technical Note [D1] has been found to be identical to the earlier NBS nomographs [D2]. The nomographs were developed to avoid the cumbersome calculations needed for deformation corrections. However, microcomputers have made the formula easier to use than nomographs, and we therefore give only the deformation formula.

$$\alpha \approx 2.231 \, P^{\frac{2}{3}} (V_1 + V_2)^{\frac{2}{3}} D^{-\frac{1}{3}} \qquad (D.1)$$

α is the deformation in millimeters

P is the force in NEWTONS,

D is the diameter of the probe in MILLIMETERS,

V_i is a number characteristic of the probe or block material from the table below:

Material	V (10^{-8})
Steel	139
Chrome Carbide	86
Tungsten Carbide	40
Ceramic	139
Diamond	43

Example 1: A steel gauge block is measured with a two point probe system each with a spherical diamond contact of 6 mm diameter. The top force is 1 N and the bottom force is 1/3 N.

Top Contact:
$D = 6$ mm
$F = 1$ N
$V_1 = 139 \times 10^{-8}$ steel block
$V_2 = 43 \times 10^{-8}$ diamond sphere

$$\alpha = 2.231 * (1)^{2/3} * (139 \times 10^{-8} + 43 \times 10^{-8})^{2/3} * (6)^{-1/3} \qquad (D.2)$$

$$= 0.000181 \text{ mm} = 0.181 \text{ μm}$$

Bottom Contact:
- D = 6 mm
- F = 1/3 N
- $V_1 = 139 \times 10^{-8}$ steel block
- $V_2 = 43 \times 10^{-8}$ diamond sphere

$$\alpha = 2.231 * (1/3)^{2/3} * (139 \times 10^{-8} + 43 \times 10^{-8})^{2/3} * (6)^{-1/3} \tag{D.3}$$

$$= 0.000087 \text{ mm} = 0.087 \text{ μm}$$

Thus the total deformation is 0.00027 mm or 0.27 μm.

Example 2: Same apparatus but measuring a chrome carbide gauge block:

Top Contact:
- D = 6 mm
- F = 1 N
- $V_1 = 86 \times 10^{-8}$ chrome carbide block
- $V_2 = 43 \times 10^{-8}$ diamond sphere

$$\alpha = 2.231 * (1)^{2/3} * (86 \times 10^{-8} + 43 \times 10^{-8})^{2/3} * (6)^{-1/3} \tag{D.4}$$

$$= 0.000145 \text{ mm} = 0.145 \text{ μm}$$

Bottom Contact:
- D = 6 mm
- F = 1/3 N
- $V_1 = 86 \times 10^{-8}$ chrome carbide block
- $V_2 = 43 \times 10^{-8}$ diamond sphere

$$\alpha = 2.231 * (1/3)^{2/3} * (86 \times 10^{-8} + 43 \times 10^{-8})^{2/3} * (6)^{-1/3} \tag{D.5}$$

$$= 0.000070 \text{ mm} = 0.070 \text{ μm}$$

Thus the total deformation is 0.00022 mm or 0.22 μm.

If we were to use a steel master block to calibrate a chrome carbide block under the above conditions we see that the penetration correction for steel and chrome carbide blocks differ substantially. The total length correction of the calibrated block is given by:

$$L_u = M_u - M_s + L_s + (\alpha_s - \alpha_u) \tag{D.6}$$

where L_u is the calibrated length of the unknown block, M_u is the measured value of the unknown block, M_s is the measured value of the reference standard, and α_s and α_u are penetration corrections of the standard and unknown respectively, and $\alpha_s - \alpha_u = 0.05$ μm (2.0 μin).

REFERENCES

[D1] Puttock, M.J. and E.G. Thwaite, "Elastic Compression of Spheres and Cylinders at Point and Line Contact," Natoinal Standards Laboratory Technical Paper No. 25, CSIRO, 1969.

[D2] Beers, J.S. and J.E. Taylor. "Contact deformation in Gauge Block Comparisons," NBS Technical Note 962, 1978.

www.ingramcontent.com/pod-product-compliance
Lightning Source LLC
Chambersburg PA
CBHW080255180526
45167CB00006B/2533